Veterinary Vaccines

Current Innovations and Future Trends

Editors:

Laurel J. Gershwin
School of Veterinary Medicine
University of California, Davis
Davis, CA
USA

Amelia R. Woolums
College of Veterinary Medicine
Mississippi State University
Mississippi State, MS
USA

Caister Academic Press

www.caister.com

ISBN: 978-1-913652-59-3 (paperback)
ISBN: 978-1-913652-60-9 (ebook)

DOI: https://doi.org/10.21775/9781913652593

Contents

Current books of interest

- Climate Change and Microbial Ecology 2020
- Alphaherpesviruses 2020
- Legionellosis Diagnosis and Control in the Genomic Era 2020
- Bacterial Viruses: Exploitation for Biocontrol and Therapeutics 2020
- Microbial Biofilms: Current Research and Practical Implications 2020
- Astrobiology: Current, Evolving and Emerging Perspectives 2020
- *Chlamydia* Biology: From Genome to Disease 2020
- Bats and Viruses: Current Research and Future Trends 2020
- SUMOylation and Ubiquitination: Current and Emerging Concepts 2019
- Avian Virology: Current Research and Future Trends 2019
- Microbial Exopolysaccharides: Current Research and Developments 2019
- Polymerase Chain Reaction: Theory and Technology 2019
- Pathogenic Streptococci: From Genomics to Systems Biology and Control 2019
- Insect Molecular Virology: Advances and Emerging Trends 2019
- Methylotrophs and Methylotroph Communities 2019
- Prions: Current Progress in Advanced Research (Second Edition) 2019
- Microbiota: Current Research and Emerging Trends 2019
- Microbial Ecology 2019
- Porcine Viruses: From Pathogenesis to Strategies for Control 2019
- *Lactobacillus* Genomics and Metabolic Engineering 2019
- Cyanobacteria: Signaling and Regulation Systems 2018
- Viruses of Microorganisms 2018
- Protozoan Parasitism: From Omics to Prevention and Control 2018
- Genes, Genetics and Transgenics for Virus Resistance in Plants 2018
- Plant-Microbe Interactions in the Rhizosphere 2018
- DNA Tumour Viruses: Virology, Pathogenesis and Vaccines 2018
- Pathogenic *Escherichia coli*: Evolution, Omics, Detection and Control 2018
- Postgraduate Handbook 2018
- Enteroviruses: Omics, Molecular Biology, and Control 2018
- Molecular Biology of Kinetoplastid Parasites 2018
- Bacterial Evasion of the Host Immune System 2017
- Illustrated Dictionary of Parasitology in the Post-Genomic Era 2017
- Next-generation Sequencing and Bioinformatics for Plant Science 2017
- Brewing Microbiology: Current Research, Omics and Microbial Ecology 2017
- Metagenomics: Current Advances and Emerging Concepts 2017
- The CRISPR/Cas System: Emerging Technology and Application 2017
- *Bacillus*: Cellular and Molecular Biology (Third edition) 2017
- Cyanobacteria: Omics and Manipulation 2017
- Foot-and-Mouth Disease Virus: Current Research and Emerging Trends 2017

www.caister.com

Chapter 1

New Approaches to Vaccinology Made Possible by Advances in Next Generation Sequencing, Bioinformatics and Protein modeling

Amelia R. Woolums* and Cypriana Swiderski

Department of Pathobiology and Population Medicine, Mississippi State University, Mississippi State, MS 39759 USA.

*Corresponding author: amelia.woolums@msstate.edu

DOI: https://doi.org/10.21775/9781913652593.01

Abstract

Vaccines can be powerful tools, but for some diseases, safe and effective vaccines have been elusive. New developments in nucleic acid sequencing, bioinformatics, and protein modeling are facilitating the discovery of previously unknown antigens through reverse vaccinology approaches. Sequencing the complementarity-determining region of antibodies and T cell receptors allows detailed assessment of the immune repertoire and identification of paratopes shared by many individuals, supporting the selection of antigens that may be broadly protective. Systems vaccinology approaches to asses the global host response to vaccination by evaluation of differentially expressed genes in blood, cellular or tissue transcriptomes can reveal previously unknown pathways and interactions related to protective immunity. While it is important to remember that discoveries made through reverse vaccinology and systems vaccinology must still be confirmed with traditional challenge models and clinical trials, these approaches can provide new perspectives that may help solve longstanding problems in veterinary vaccinology.

Introduction

At its best, vaccination has substantial beneficial impact. The history of vaccination includes notable successes, such as the eradication of smallpox in humans and rinderpest in cattle (Greenwood, 2014). In ongoing campaigns, vaccines are used to prevent deaths due to pathogens such as rabies virus (Fisher and Schnell, 2018), and morbidity associated with agents such as foot-and-mouth disease virus (Diaz-San Segundo *et al.,* 2017). It is indisputable that vaccines can be powerful tools in the promotion of animal and human health. However, it has been difficult to develop reliable vaccines for some diseases, and vaccination has even rarely enhanced disease (Kapikian *et al.,* 1969). Clearly, information required to guide development of safe and effective vaccines is sometimes lacking.

In situations where knowledge gaps impede vaccine development, approaches leveraging new technologies that allow relatively rapid and inexpensive determination of nucleic acid sequences or protein structure may provide the missing insights. Here we summarize recent research using such technologies to identify new antigens, to characterize more holistically protective host responses to antigens and adjuvants, and to investigate adverse vaccine reactions. While much of the available information to date comes from investigations of human vaccines, the examples demonstrate the possibilities for answering a variety of questions relevant to veterinary vaccinology.

Vaccine development: classical versus next generation

Classical vaccinology is a hypothesis-driven approach to laboratory testing of pathogens that identifies antigenic components capable of eliciting protective immunity (Moxon *et al.,* 2019). Classical vaccinology employs a pipeline that begins by isolating the pathogen, followed by some manner of attenuation that retains immunogenicity, stimulates memory, and prevents pathogenicity. Subsequent iterations of this approach have applied biochemical, serological and microbiological methods to purify antigenic factors from both attenuated organisms and from organisms grown in culture— a laborious approach that yields few immunogens over very protracted time frames (Rappuoli, 2000). Clearly, classical vaccinology approaches have been informed by insight garnered from natural infections, such as the use of immune sera for screening candidate antigens, and recognition that farmers previously infected with cowpox did not develop lesions following smallpox variolization, nearly 28 years prior to Jenner's application of this principle in the first vaccination (Boylston, 2013). Nonetheless, classical vaccinology relies heavily on injection into disease surrogates (most commonly laboratory animals) in order to identify the antigenicity of vaccine candidates and their ability to elicit a protective immune response.

Contrasting classical vaccinology, scientific and technological advances have ushered in a new generation of vaccinology in which candidate vaccine antigens are computationally predicted from the DNA sequences of pathogen genomes (Capecchi *et al.,* 2004). Moreover, the ability of peptides from these proteins to bind MHC molecules and become efficient T-cell epitopes; or to act as linear, conformational, or immunoglobulin class-specific (Saravanan and Gautham, 2018; Gupta *et al.*, 2013) B-cell epitopes, is being predicted *in silico* (reviewed in Dhanda *et al.,* 2017). Quite significantly, antigen-specific sequences of B-cell and T-cell receptors (ie paratopes) derived from individuals with protective immune responses are being characterized to identify their complementary antigenic epitopes. Paralleling these technological advances in vaccine antigen development are co-evolving technologies that are increasing knowledge and understanding of the mechanistic basis of protective immunity in the context of specific pathogens, as will be discussed later in this paper.

By refining candidate vaccine antigen prediction and the discernment of immune effector pathways that confer protective immunity, next generation vaccinology improves the efficiency with which animals are used to screen vaccine candidates for immunogenicity and protective immune responses. Significantly, the *in silico* predictive approaches, such as those used for candidate vaccine antigen prediction, reflect algorithms that are constantly evolving. Subsequent algorithms predict antigens that are both conserved, and novel relative to prior algorithms, while also failing to identify previously predicted antigens (Dalsass *et al.,* 2019). These discrepancies, which are inherent in the approach, highlight 1) a need for redundancy in the computational predictions employed to identify candidate vaccine antigens, and 2) the importance of comprehensive validation of the ability of predicted antigens to elicit protective immune responses in pre-clinical models and ultimately in the host species for the pathogen.

Reverse Vaccinology

Classical vaccinology successes are readily demonstrated for pathogens that lack antigenic variability or for which immunologic memory prevents reinfection (Rappuoli, 2007). However, most pathogens do not possess these characteristics. Classical vaccinology has proven inadequate for the development of vaccines against pathogens demonstrating antigenic diversity, pathogens that cannot be grown in vitro or lack adequate animal models of infection, and pathogens for which cell-mediated immune responses are protective (Rappuoli, 2000). Certain vaccines have also demonstrated safety issues tracing to the potential for recombination and immunologic events that result in enhanced disease (Agnew-Crumpton *et al.,* 2016, Kapikian *et al.,* 1969).

Contrasting the need for pathogen isolation with classical vaccinology, the advent of massively parallel DNA sequencing, also termed next generation sequencing (NGS), enabled rapid, cost-effective determination of a pathogen's genome sequence. Using a methodological approach coined *reverse vaccinology* by Rappuoli (2000), the full repertoire of protein coding sequences (the pathogen's proteome) is identified from the genome sequence of the pathogen. For bacteria and more complex pathogens, this step yields thousands of proteins, the majority of which are not relevant as vaccine antigens. To narrow the predicted proteins to those most relevant as candidate vaccine antigens, an expanding number of computational algorithms have been developed that model the encoded proteins from their DNA sequences, in order to predict characteristics of the proteins such as their 3-dimensional structure and predicted expression sites in the pathogen. Using this approach, for instance, proteins predicted to be secreted or exposed on the surface of the pathogen in particular conformations, and therefore subject to immune surveillance, can be identified and prioritized for analysis as vaccine candidates. Following this *in silico* prediction, DNA sequences of these candidate vaccine antigens are compared to genomic sequences of the host and homologous sequences with the potential to serve as autoantigens are eliminated.

Reverse vaccinology was first applied to Group B *Neisseria meningitidis,* an important cause of bacterial meningitis in humans. Whereas vaccines for other groups of *Neisseria meningitidis* employ the bacteria's capsular polysaccharide as an antigen, this approach is complicated by conserved polysialic acid residues in the bacteria's capsular polysaccharide that are shared with human tissues (Serruto *et al.,* 2012). Employing the genome sequence of Group B *Neisseria meningitidis* (strain MC58), more than 2000 predicted proteins were analyzed *in silico* to identify proteins predicted to be surface exposed or excreted (Pizza *et al.,* 2000). Of the 570 predicted proteins, 350 were successfully expressed, purified, and used to immunize mice. Immune sera were screened by Western blot analysis against bacterial extracts to verify protein expression, while surface expression of the protein was confirmed by enzyme-linked immunosorbent assay (ELISA) and flow cytometry using intact, whole bacteria. Finally, complement-mediated killing activity of the antibodies (using human complement) was evaluated because it is an accepted correlate for in vivo protection in clinical trials of human meningococcal vaccines (Borrow *et al.,* 2006). Of the 91 proteins found to be positive in at least one of the first three assays, 28 induced antibodies with bactericidal activity (Serruto *et al.,* 2012). Five proteins were included in the final vaccine based on their ability to induce protection against diverse *N. meningitidis* strains. This protection was assessed by determining if specific antibodies to each protein antigen conferred passive protection in infant rodent models, or by identifying serum bactericidal antibodies following vaccination (Giuliani *et al.,* 2006). The final vaccine (4CMenB, Bexsero®) was released in Europe in 2013 (and the United States in 2015), 13 years following publication of the pathogen's genome sequence and initiation of the vaccine effort (Tettelin *et al.,* 2000).

While reverse vaccinology presents distinct advantages resulting from identification of the complete protein repertoire of a pathogen, dependence on protein coding sequences prevents identification of non-protein antigens such as polysaccharides, which have been important components of many successful vaccines, and CD1-restricted glycolipids which are promising vaccine candidates (Rappuoli, 2000). Also noteworthy is the finding that certain proteins can be restrictive in terms of their pathogen recognition and poorly antigenic (Sundling *et al.,* 2013), challenges that are circumvented by the use of adjuvants and construction of multi-epitope vaccines (Burton, 2017).

In Silico Prediction of B-cell Receptor/Antibody Epitopes

The large number of candidate protein antigens identified by reverse vaccinology presents a significant bottleneck in the screening of these antigens to determine their immunogenicity. Purified, full-length recombinant proteins used to elicit antibodies are often difficult to prepare in quantities sufficient for immunization. This is exemplified by the aforementioned screening of 570 predicted candidate vaccine proteins, 350 of which could be expressed in developing the Group B *N. meningitidis* vaccine, 4CMenB. To decrease the numbers of proteins subjected to *in vivo* screening, downstream technologies have been developed to increase the likelihood that predicted proteins from the genome sequence of a pathogen will be immunogenic. These analyses to predict immunogenic epitopes reflect the field of immunoinformatics that utilizes bioinformatics approaches to understand and interpreting immunological data.

B-cells are well recognized as the source of antibodies that provide protection from pathogens and cancerous cells. Antibodies recognize their target antigen by binding to portions of the antigen that are termed antigenic determinants or epitopes. Antigens generally possess many sites that function as epitopes. Regions of the antibody to which epitopes bind are termed paratopes, which are composed of six complementarity-determining regions (CDR) that confer antigen-specificity in epitope to paratope binding. Immunologic dogma holds that an antibody binds to a single epitope on an antigen, but expanding evidence indicates that some antibodies bind to more than one epitope (Van Regenmortel, 2014). A B-cell (and its clones) secretes, and also expresses on its surface, antibodies to the same epitope. Thus, antibodies expressed on the B-cell surface function as receptors (B-cell receptors) for the same epitope as the secreted antibodies. B-cell receptor stimulation by B-cell epitopes contributes to both the development of immunologic memory and antibody secretion. Accordingly, B-cell epitope prediction is an important goal to improve the likelihood that candidate vaccine proteins in reverse vaccinology are immunogenic.

B-cell epitopes can be linear (continuous) or conformational (discontinuous) (Stave and Lindpainter, 2013). Linear B-cell epitopes are formed by sequential amino acids in the protein. Conformational B-cell epitopes are formed by amino acids that are not sequential in the protein, but instead come into close contact to form an epitope as a result of the three-dimensional shape of the folded protein. Tools for *in silico* B-cell epitope have been reviewed elsewhere (Dhanda *et al.,* 2017). A primary challenge for *in silico* B-cell epitope prediction is that the majority of B-cell epitopes have been shown to be conformational (Greenbaum *et al.,* 2007; Kringelum *et al.,* 2013;

Ferdous *et al.,* 2019;). Illustrating the complexity of this challenge, Stave and Lindpaintner (2013) evaluated crystal structures from 111 antigen-antibody structures derived from proteins 22-442 amino acids long. They identified discontinuous epitopes that were formed from peptides ranging in length from 20 to 442 amino acids (average 50-79 amino acids) in which 20-101 (median=37) residues were in contact between the epitope and paratope, in clusters of 2-12 amino acids. Indeed, *in silico* B-cell epitope prediction tools that are based upon protein sequence attributes including calculations of hydrophilicity, flexibility, beta-turns, surface accessibility, amino acid composition and amino acid cooperativeness have been poorly predictive (Kringelum *et al.,* 2012; Jespersen *et al.,* 2017). Given that the theoretical B-cell receptor repertoire is nearly unlimited (10^{74} possible sequences, Saada *et al.,* 2007), the likelihood that any surface accessible region of a protein will have a complementary antibody conformation is rather high. Accordingly, the poor predictive value of *in silico* B-cell epitope predictions is likely an inherent characteristic of tools that are designed to comprehensively identify epitopes that are 'potentially' antigenic (Jespersen *et al.,* 2019). Supporting this assertion, conformational B-cell epitope predictions employing various artificial intelligence platforms that are trained in a more supervised manner, using crystal epitope structures that correspond to well characterized antibody paratopes, provide improved epitope predictions (Dhanda *et al.,* 2017; Jespersen *et al.,* 2017). However, these tools employ resolved crystal structures of target antigens that are not generally available. Addressing and perhaps solving this conundrum, Rahman *et al.* (2016) demonstrated, using a B-cell epitope prediction tool that characterizes sequence-based protein disorder tendency (IUpred-L), that epitope prediction based on short ≤11-aa peptides falsely classifies B-cell epitopes as non-epitopes (53% accuracy). This is because short peptides of B-cell epitopes bind poorly. By modeling peptides of moderate length (16-30 amino acids) epitope prediction achieved 86% accuracy, indicating that training sets composed of appropriately sized longer peptides are necessary for accurate *in silico* B-cell epitope prediction.

In Silico Prediction of T-cell Receptor Epitopes

Cell-mediated immunity is mediated by T-cells that have cytotoxic activity, ie cytotoxic T-cells, or by helper T-cells that provide co-stimulatory signals for B-cells and cytotoxic T-cells. Both helper T-cells and cytotoxic T-cells recognize epitopes in a highly specific manner, due to the antigen-specificity of paratopes in their T cell receptors (TCRs). However, unlike B-cell receptors, TCRs recognize epitopes that are bound to Major Histocompatibility (MHC) receptors— specifically MHC I for cytotoxic T-cells, and MHC II for helper T-cells. MHC I receptors are expressed on the surface of all nucleated cells whereas MHC II receptors are found on antigen presenting cells. MHC genomic loci are among the most variable genes in mammals, enabling great diversity in MHC molecules. In humans, where MHC expression has been well characterized, there are at least 10,000 MHC I alleles (Robinson *et al.,* 2017) and 3,000 MHC II alleles (Rock *et al.,* 2016). Individuals express 3-6 different MHC I alleles (3 from each parent) and 3-12 different MHC II receptor alleles (Rock *et al.,* 2016). Each MHC allele is estimated to be able to bind 20 million (MHC I) to 200 billion (MHC II) epitopes (Rock *et al.,* 2016). Interaction between the T-cell receptor paratope and peptide bound to an MHC receptor confers antigen-specific recognition. A primary challenge for *in silico* TCR epitope prediction is evident in the possible combinations that are generated when MHC diversity is considered with the potential diversity in the antigen recognition regions of TCRs (ie TCR paratopes).

Approaches to *in silico* T-cell epitope prediction can be direct or indirect. Direct prediction targets identification of epitopes that bind to TCRs, while indirect prediction targets identification of epitopes that bind to MHC receptors. Indirect T-cell receptor epitope prediction is based on peptide binding to a groove within both MHC I and MHC II molecules in which peptides are maintained predominantly in an extended conformation (Madden, 1995). Early direct prediction algorithms were based upon structural analyses of helper T-cell epitopes that indicated these epitopes were amphipathic with helical turns (DeLisi and Berzofsky, 1985; Margalit *et al.,* 1987; Stille *et al.,* 1987). Subsequently, these finding were extended to cytotoxic T-cell epitopes (Reyes *et al.,* 1988). However, additional investigations of the MHC II peptide binding groove demonstrated that it cannot accommodate helices, but instead binds to linear peptides (Stern *et al.,* 1994).

MHC I molecules load peptides generated by proteasomal proteolysis, binding to the peptides in the endoplasmic reticulum (ER), after the peptides are translocated from the cytosol (Blum, *et al.,* 2013). The peptide-binding groove of MHC I molecules is closed at both ends and accommodates peptides that are typically 8 to 11 amino acids long (Rammensee *et al.,* 1995). N- and C-terminal ends of the peptide form hydrogen bonds with amino acids that are conserved across MHC I molecules, and the peptide-binding groove contains deep binding pockets with tight physicochemical preferences (Madden, 1995). Collectively, these features enable binding predictions. MHC II bound epitopes classically result from cleavage of pathogen proteins to peptides in the endolysosome of antigen presenting cells, followed by association of peptide epitopes with MHC II molecules exiting the Golgi. Peptides bind to an MHC II binding groove through a series of hydrogen bonds in sites that are highly conserved within the MHC II molecule, despite differences in peptide epitope sequences and MHC proteins (Stern *et al.,* 1994). This confers a stereotyped but complex mode of binding across the spectrum of peptide-MHC II interactions that informs *in silico* prediction algorithms for MHC II binding epitopes. An important distinction between MHC I and MHC II molecules is that the binding groove for MHC class II is open at both ends (Painter and Stern, 2012). The option for bound peptide to protrude from the MHC molecule makes *in silico* MHC II binding prediction more difficult (Lundegaard *et al.,* 2010). As a consequence of the extension beyond the groove, peptides that bind MHC class II molecules tend to be of variable length but are typically 12 to 25 amino acids long (Jardetzky *et al.,* 1996).

Computational tools that predict peptide binding to MCH molecules are predicated on the availability of high-quality DNA sequence data for MHC alleles that are highly prevalent in the population that will be vaccinated. In humans >22,000 human MHC allele sequences have been deposited into the Immuo Polymorphism MHC Database (www.ebi.ac.uk/ipd/mhc/). In contrast, there are significantly fewer classical MHC alleles in this database for the horse, sheep, dog, pig and cow: currently 60, 247, 384, 450 and 678, respectively. MHC allele expression is vastly different across human ethnic groups (Terasaki, 2007), limiting world-wide application of T-cell epitope-based vaccination. This is relevant for veterinary medicine where the diversity and prevalence of MHC alleles within each species tends to be poorly characterized. Vaccines derived from MHC-peptide binding predictions are efficacious in individuals that express the MHC allele used for the epitope prediction. In pigs and cattle, where MHC allele expression is more likely to be defined within large portions of the population, these prediction methods are being applied. In this regard, MHC alleles have been identified in humans and cattle that share a high degree of epitope specificity (Lund *et al.,* 2004; Pandya *et al.,* 2015). These MCH supertypes enable T-cell epitope vaccines to address large populations with diverse MHC expression based upon expression of at least one allele in the supertype. Accordingly, prioritizing the identification of MHC supertypes in species with highly polymorphic MHC molecules has particular advantages for vaccine design.

Structure based methods are available for predicting peptide binding to MHC complexes but these methods have low predictive performance (Nielsen *et al.,* 2018). In contrast, matrix-based predictions and machine learning technologies that employ protein sequences as input and are trained on known MHC-peptide interactions have improved predictive performance (Mei *et al.,* 2019; Nielsen *et al.,* 2018). Few of these tools have been validated on veterinary species. However, NetMHCpan employs artificial neural networks that have been trained through multiple iterations on peptide-MHC I affinity measurements from human, mouse, primates, cattle, and swine, as well as ligands and their complementary MHC I alleles from these species (Hoof *et al.,* 2009; Carrasco Pro *et al.,* 2014; Jurtz *et al.,* 2017; Hansen *et al.,* 2014). NetMHCpan can predict peptide-MHC class I binding for any allele of known sequence. Nine of 19 MHC I binding epitopes predicted by this tool for foot and mouth disease virus (FMDV, >1400 possible peptides) bound to pig MHC I (i.e. SLA) (Pedersen *et al.,* 2013). A web interface of NetMHCpan that is specifically trained for identifying MHC I restricted peptides in cattle is available (Nielsen *et al.,* 2018). This tool has been used to identify MHC I-peptide binding for FMDV (Pandya *et al.,* 2015), bovine herpes virus, and *Theileria parva*, the causative agent of East Coast Fever (Nielsen *et al.,* 2018; Svitek *et al.,* 2014). An alternative proprietary prediction tool for MHC I and MHC II peptide binding, PigMatrix, is based on the pocket method of MHC I-peptide binding, and has been used to identify T-cell vaccine epitopes to for influenza virus (Gutiérrez *et al.,* 2015; Hewitt *et al.,* 2019).

Additional Antigen Selection Criteria Employed in Reverse Vaccinology

RNA-sequencing (RNAseq) is a technology that both identifies and quantifies RNA sequences that are expressed within a biological sample. In one embodiment of its utility for candidate vaccine antigen identification, RNAseq has been used to quantify the level to which *in silico* predicted vaccine candidates are expressed (as mRNA) by the pathogen. This can be done for instance during conditions of natural infection, or during surrogates of natural infection such as low pH or osmotic stress, as would be anticipated in the phagosome. These results are then informative in ranking the *in silico* predicted vaccine candidates based upon the relevance of their expression levels. Similarly, proteomic analysis, which identifies the proteins that are expressed and their magnitude of expression can also be employed to quantify expression of putative vaccine peptide epitopes in order to assist in ranking their relevance.

Reverse Vaccinology for Veterinary Pathogens

Employing a pathogen's genome sequence to predict vaccine epitopes using reverse vaccinology presents distinct advantages over classical approaches. This approach has enabled vaccine design in for problematic pathogens that cannot be cultivated. Such is the case for *Pajaroellobacter abortibovis,* the causative agent of foothills abortion in cattle, which can only be cultivated in SCID mice (Blanchard *et al.,* 2010). Using a reverse vaccinology approach, Welly *et al.* (2017) first determined the genome sequence of the pathogen by sequencing DNA from infected SCID mice. Computational methods were used to subtract the well-characterized genomic sequences originating from the SCID mice, leaving the remaining genomic regions belonging to *P. abortibovis*. DNA regions belonging to *P. abortibovis* were then arranged into a complete *de novo* genome sequence from which the authors identified 10 putative vaccine candidates using *in silico* prediction. Seven of these predicted proteins, when expressed as fusion proteins, were recognized by serum from *P. abortibovis* infected SCID mice, indicating that the proteins could be relevant to protective immunity.

In other research, a reverse vaccinology approach was recently used to identify cat flea (*Ctenocephalides felis*) surface antigens predicted to be immunogenic. When the antigens were incorporated into experimental vaccines given to cats, flea eggs collected from vaccinated cats were significantly less likely to hatch ($P < 0.05$), as compared to flea eggs collected from cats receiving adjuvant only (Contreras *et al.*, 2018). Tomazic *et al.* (2018) used reverse vaccinology to identify antigens of *Cryptosporidium parvum* that reacted with antibodies in serum collected from *Cryptosporidium*-infected calves in the first month of life, indicating that the antigens could be associated with protection at the time when calves are most vulnerable to diarrhea resulting from cryptosporidiosis. In both of these examples reverse vaccinology allowed rapid screening of a large number of candidate proteins to identify antigens which might have been overlooked in classical approaches using whole pathogens or their components to induce immunity with experimental challenge.

Reverse vaccinology also presents advantages when dealing with pathogens that exhibit antigenic diversity. For instance, the DNA from multiple strains of a given bacterial species, termed a pangenome, can be simultaneously analyzed to allow *in silico* prediction of shared epitopes. This approach was used to identify T cell epitopes in antigens shared among 30 *Acinetobacter baumannii* strains, with a goal to developing effective vaccines for this bacterial pathogen that demonstrates widespread resistance to antimicrobials (Hassan A *et al.*, 2016). Also relevant is the case of many RNA viruses where host selection pressure drives gene mutations that contribute to antigenic diversity. For example, sequencing of variable regions in nonstructural proteins 1 and 2 (nsp1 and nsp2) and the products of open reading frame 3 and 5 (ORF3 and ORF5) in pigs infected with porcine reproductive and respiratory syndrome virus (PRRSV) revealed genetic changes related to virus rebound at 42 days post infection (Chen *et al.*, 2016). Improved understanding of the mechanisms by which PRRSV escapes host immunity could facilitate development of vaccines that more effectively prevent infection.

Epitope-Focused Vaccine Design

Whereas massively parallel DNA sequencing provided a gateway to reverse vaccinology, it is now possible to identify the breadth of the immune repertoire by sequencing DNA encoding antibody and T-cell receptors (TCR) from hundreds of thousands of individual immune cells (single cell sequencing) (Davis and Boyd, 2019). Further, sequencing and quantification of RNA sequences that encode antibody and TCRs from large populations of immune cells can be used to identify changes in immune repertoire expression during responses to vaccination (DeKosky *et al.*, 2013; Waickman *et al.*, 2019) or infection (Sant *et al.*, 2018; Mitsunaga and Synder, 2020). Epitope-focused vaccine design (subsequently coined Reverse Vaccinology 2.0 (Rappuoli *et al.*, 2016), refers to the design of vaccine epitopes that bind to paratopes on an antibody or on a T-cell receptor that are known to confer protection during infection (Correia *et al.*, 2014; Sela-Culang *et al.*, 2015). Prediction of complementarity between an antigen epitope and an antibody paratope differs significantly from predicting B-cell epitopes. This is because prediction of epitope/paratope complementarity is constrained by stringent epitope/paratope interactions, while only surface accessibility constrains B-cell epitope prediction and accordingly yields a large number of protein sequences (Novotný *et al.*, 1986).

Epitope focused vaccine design can be viewed in four stages, with different methodological approaches being available to address each stage. First, individuals are identified that have protective immune responses to a pathogen, such as broadly neutralizing antibodies, bactericidal antibodies, or antigen-specific helper T-cell or cytotoxic T-cell responses. Second, B-cells or T-cells with paratopes that confer the desired immunologic outcome are isolated and the paratope sequences are determined. Third, structural features of the cloned paratopes are computationally modeled in order to predict the structure of the corresponding antigen epitope. Fourth, immunological testing and validation of the predicted antigen epitope is undertaken. In one embodiment of this approach, antibodies providing the majority of virus neutralizing activity for human respiratory syncytial virus were critical to the identification of an antigenic but transient fusion protein conformation. By inserting point mutations that stabilized the fusion protein structure, the antigen has been advanced as a vaccine candidate that has been effective in clinical trials (Magro *et al.*, 2012; Gilman *et al.*, 2016; Crank *et al.*, 2019). Alternatively, *in silico* co-modeling of CD4+T-cell receptor and MHC II sequences from human patients with latent tuberculosis, who were vaccinated with *M. tuberculosis* peptides, identified vaccine epitopes that triggered CD4+ T-cell responses (Dash *et al.*, 2017). While this approach has theoretical strength for identifying key protective antigens, its primary drawback is the resolution of only a portion of the full repertoire of epitope-specific antibodies and T-cell receptors, which limits comprehensive characterization of antigen-specific immune effector mechanisms.

Assessment of the host response: "systems vaccinology"

An effective immune response is the result of the choreographed interaction of thousands of molecular or cellular reactions which may be separated in the host by the distance of a meter or more. As the number of recognized components participating in the immune response has grown, it has become obvious that historical approaches to measuring immunity, which often focused only on serum antibody or a few cellular reactions, missed much of the picture. While for some applications one or a few signature responses may provide enough information, in cases where the nature or mechanics of a protective response are unclear, a more detailed portrait may be informative. This is the premise of "systems vaccinology":

development and assessment of vaccines with more of the immune response in view.

Systems vaccinology can be considered a branch of systems biology, in which interdisciplinary approaches are used to characterize the complex networks within a biological system, in order to better understand and predict the behavior of the system (Pulendran *et al.,* 2010). A systems biology assessment evaluates all parts of a system responding to some perturbation, analyzing the results together in order to develop a model that more accurately represents the nature of the whole. Systems vaccinology approaches can reveal the involvement of formerly unknown genes and pathways related to the response to a vaccine; thus, these methods are being applied to identify correlates that predict protective immunity, to reveal mechanisms of vaccine or adjuvant efficacy, and to uncover factors contributing to adverse vaccine reactions.

In spite of the potential of systems vaccinology, the use of this approach may be limited by the high cost of RNAseq or other required technologies, which may substantially limit the number of individuals that can be evaluated, thus weakening study power. The complications of analyzing massive data sets is also a formidable hurdle. It is also important to remember that the findings of a systems vaccinology investigation must still be confirmed with prospective investigations requiring use of more classical methods. Therefore, systems vaccinology should be considered a complement to traditional vaccinology, and not a replacement.

Predicting protective immunity induced by vaccination

While some vaccines are reliably protective, many are not. Moreover, the correlates of protective immunity are known and can be measured for some vaccines, but they are not clear for others. And, at times, immune responses that are assumed to be important are not identified in vaccinated patients that are nevertheless protected. In cases such as these, a global assessment of the immune response may help to reveal unknown but important mechanisms. In most cases to date this global systems vaccinology assessment has been made by assessing gene expression in blood or other tissue, by measuring hundreds or thousands of transcripts, most commonly using microarrays or, more recently, high throughput RNAseq. In humans and livestock, blood or peripheral blood mononuclear cell (PBMC) transcriptomes have been assessed, because blood is usually easy to collect, and also because lymphocytes and antigen presenting cells traffic through blood during their response to vaccination at various sites. Increased gene expression over baseline in whole blood or PBMC can sometimes be seen within 24 hours of vaccination (Vahey *et al.,* 2010), with expression at 24 hours sometimes significantly associated with outcomes weeks later (Matthijs *et al.,* 2019). Transcriptomes in tissues collected from vaccinated animals subjected to necropsy at various times after vaccination or challenge have also been used to investigate vaccine responses (Luo *et al.,* 2014; Li *et al.,* 2020).

An early application of systems vaccinology was used to characterize the response of humans vaccinated with the yellow fever vaccine YF-17D (Querec *et al.*, 2009). While YF-17D has been administered for decades and is highly effective, the reasons for its efficacy are still not clear. To improve understanding of the immunologic mechanisms elicited by YF-17D vaccination, Querec *et al.* investigated post-vaccination changes in plasma cytokines, chemokines, blood antigen presenting cells, and expression of 20,000 genes in peripheral blood mononuclear cells (PBMC) by microarray over the first 21 days after vaccination in 15 individuals. These early events were compared to correlates of later protective immunity: SN titers and CD8+ (cytotoxic T cell) activation at days 15 and 60 after vaccination. Following vaccination, early events including increases in CXCL10 (previously IP-10), IL-1α, and the percent of circulating CD86+ antigen presenting cells were identified, but these events were not correlated with CD8+ responses at 60 days post vaccination. In contrast, when 839 genes that correlated to the magnitude of day 60 CD8+ responses were subjected to unsupervised principal component analysis (PCA) followed by discriminant analysis using mixed integer programming (DAMIP), predictive relationships between PBMC gene expression and day 60 CD8+ responses were identified. PCA is a statistical procedure commonly employed in the analysis of gene expression datasets which is sensitive to strong patterns of expression that are shared across samples. PCA uses an orthogonal transformation to convert expression data into a set of values of linearly uncorrelated variables called “principal components”. DAMIP is a predictive modeling framework that uses a supervised-learning classification approach used to predict biomedical phenomena. When the investigators repeated this analytic approach in a second group of vaccinated individuals, the results predicted day 60 CD8+ responses with 87%

accuracy. Similarly, the investigators were able to predict SN responses in the second group of vaccinates with 80% accuracy.

In a greatly expanded effort, this group went on to evaluate the relationship between the PBMC transcriptome in the week after vaccination and protective immunity induced by each of five vaccines: two meningococcal vaccines, YF-17D, a trivalent inactivated influenza vaccine, or a live-attenuated influenza vaccine (Li *et al.,* 2014). For each vaccine, established correlates of immunity (such as serum antibody titers) were measured in vaccinated human subjects, and correlations with genes that were differentially expressed between baseline and the week after vaccination were assessed. For subjects vaccinated with one of the two meninogococcal vaccines, 1,150 genes were differentially expressed between the day of vaccination and day 7 postvaccination. Importantly, the research showed that evaluation of the correlation between *individual* differentially expressed genes (DEG) and protective antibody titers led to few significant and meaningful relationships. However, when DEG were grouped into blood transcription modules (BTM), based on shared functions and previously described interactions, BTM were significantly associated with protective responses to all five of the vaccines evaluated. Notably, some BTM significantly associated with protective responses to viral vaccines were different from the BTM significantly associated with protective responses to bacterial vaccines. It seems likely that assessment of gene expression by evaluating genes together in functional groups or pathways, rather than individually, may provide a more accurate picture of the host response to vaccination. Also, evaluating the response of groups of genes with coordinated expression may reveal significant changes when change in expression of any one gene may be too small to detect (Haining 2014).

The BTM developed by Li et al. were later used to evaluate early immunologic events in pigs responding to five different *Mycoplasma hyopneumoniae* vaccines (Matthijs *et al.*, 2019). These investigators identified an early upregulation of modules associated with innate immunity, including monocyte and neutrophil function, as well as inflammation and pathogen sensing within 24 hours of vaccination, which correlated to adaptive immune responses to *M. hyopneumoniae* at later time points. Significantly, pigs demonstrating increased expression of genes in these innate immune modules more than 24 hours following vaccination had weaker adaptive responses than early responding pigs. Collectively these findings indicate that the best adaptive immune response to *M. hyopneumoniae* vaccination was associated with increased gene expression that supports innate immunity in the first 24 hours after vaccination, followed immediately by decreased expression.

Researchers have subsequently used similar methods to characterize the response to other veterinary vaccines, for example, in German Landrace pigs and Pietrain pigs vaccinated against porcine respiratory and reproductive syndrome virus (PRRSV; Islam *et al.* 2019), in carp vaccinated orally with a DNA vaccine against *Vibrio mimicus* (Li *et al.*, 2020), and in horses vaccinated against African horsesickness virus (Pretorius *et al.*, 2016). However, to date relatively few groups have used blood transcriptome assessment developed in one cohort to predict vaccine efficacy in another cohort. Accurate and rapid prediction of vaccine efficacy in populations, without the need to wait to assess response to disease challenge, is a major goal of systems vaccinology. Ideally, findings initially made in relatively small groups with expensive and complicated transcriptomics could be adapted to a rapid, inexpensive test, perhaps based on PCR or sequencing of a small number of transcripts, that could be applied to larger populations.

More recently, whole blood and PBMC transcriptome data has been used to identify gene expression signatures that predict antibody responses in young (<35 years) and aged (>60 years) human cohorts following influenza vaccination in different regions of the United States over multiple years (HIPC-CHI Signatures Project Team and HIPC-I Consortium, 2017). These investigators identified high and low antibody responders based on changes in their serum antibody titers (hemagglutination inhibition or SN) following vaccination. By evaluating changes in the expression of more than 32,000 genes in response to vaccination, a "response score" based on expression levels of 9 differentially expressed genes (DEG) or 3 DEG modules (similar to the BTM of Li *et al.*, 2014) was identified which differentiated high versus low antibody responders in young but not aged cohorts. Significantly, the response score accurately predicted individuals that were high or low antibody responders when applied to a separate young vaccination cohort. While these investigators successfully used blood transcriptome data from one cohort of vaccinates to predict protective responses in a second cohort, their work also illuminated limitations of the systems vaccinology approach: 1) multiple cohorts were needed to identify the DEG that predicted SN antibody responses because there was too much variability in any one cohort to find this relationship; and 2) the validity of the findings was dependent on the population structure of the discovery cohort. Specifically, the DEG that predicted SN antibody responses were valid for individuals under 35 years of age, but not for individuals over 60 years of age. In fact, there was an inverse relationship between gene expression associated with SN responses in young individuals versus older individuals, indicating that gene expression signatures that were *positively* associated with SN antibody titers following vaccination in young individuals were *negatively* associated with SN antibody titers following vaccination in old individuals. While this is likely an important finding, the results indicate that it may not be possible to extrapolate gene expression profiles from one type of population to another. Moreover, it may be necessary to assess blood transcriptomes in large numbers of individuals, or in multiple smaller cohorts, to identify meaningful predictive signatures. Though achieving populations of this magnitude may limit similar applications in veterinary research, advances in technology that have decreased the costs of assessing a transcriptome are likely to make these applications increasingly accessible.

Mechanisms of vaccine and adjuvant efficacy

In addition to identifying genes, gene modules, or pathways of molecular interactions that predict vaccine efficacy, systems vaccinology approaches can reveal previously unknown components or mechanisms related to vaccine or adjuvant efficacy, or immunity more generally. When differential expression of dozens or hundreds of blood or PBMC transcripts is associated with a favorable response to a vaccine, some of these transcripts may be from genes that are poorly characterized, or perhaps unexpected. Further investigation of these genes in association with more classical methods to assess immunity and disease resistance can lead to discovery of relevant new mechanisms (Hagen and Pulendran, 2019). For example, the finding that expression of Toll-like receptor 5 (TLR5, a pathogen recognition receptor that binds bacterial flagellin) was associated with influenza antibody titers a month after vaccination (Nakaya *et al.*, 2011) directed further research leading to the discovery that flagellin activates production of B-cell growth factors by macrophages. Moreover, treatment of mice with antimicrobials to decrease gut bacteria (a large source of flagellin exposure for local immune cells) led to decreased antibody production following vaccination with inactivated influenza vaccine, but not modified live vaccine (Oh *et al.*, 2014). Collectively, these findings indicate that the normal intestinal flora may act as a sort of “endogenous adjuvant” for individuals vaccinated with certain vaccines. Further research in this area may lead to ways to improve response to vaccines by modifying the normal intestinal flora, or perhaps by co-treatment with probiotics. These findings also illustrate how investigation of an unexpected molecular interaction identified by a systems vaccinology approach led to discovery of a new concept in immune system regulation that is directly relevant to vaccine efficacy.

Though adjuvants have been used for over a century to improve response to vaccines, the exact immune mechanisms by which adjuvants exert their effects is surprisingly ill-defined. Systems vaccinology approaches present an opportunity to identify the genes and molecular pathways responsible for adjuvant effects. This approach has been described for adjuvants containing different cationic lipids or Toll-like receptor ligands in experimental *M. hyopneumoniae* vaccines (Mattijis *et al.,* 2019). In these experiments, cationic lipid adjuvants, which were associated with the highest serum antibody titers post vaccination, induced rapid but transient upregulation (generally limited to the first day after vaccination) in blood transcription modules related to myeloid cell activation in vaccinated pigs.

Investigation of adverse vaccine reactions

Just as systems vaccinology approaches can be used to reveal genes or pathways associated with the response to a vaccine or adjuvant, they can also be used to dissect adverse vaccine responses, in order to prevent them (Gonzalez-Dias *et al.,* 2020). To investigate the mechanistic basis of bovine neonatal pancytopenia, an adverse reaction linked to cattle vaccinated with inactivated bovine viral diarrhea virus (BVDV) vaccines, Demasius *et al.* (2013) analyzed whole blood transcriptome of vaccinated cattle. Perhaps surprisingly, the analysis identified evidence of a coordinated response to double stranded RNA, but not, as expected, to alloantigens. The investigation also led to the discovery of a gene for an apparently novel cytokine, which was strongly upregulated in vaccinated cattle. Together these findings provided evidence of unexpected aspects of the pathogenesis of neonatal pancytopenia, although further research will be necessary to establish the significance of these findings.

Systems vaccinology, summary

The immune response to a vaccine is the result of coordinated interaction of thousands of events. Given this situation, it is perhaps not surprising that research focused on a handful of molecules or cells sometimes fails to explain why a vaccine induces--or fails to induce--protection. Global assessment of gene activation in blood or other tissues through assessment of the transcriptome can provide new insights regarding mechanisms of vaccine efficacy. While reports to date have rarely combined assessment of the transcriptome with protein and metabolite expression (the "proteome" and the "metabolome"), as technology progresses and costs decrease, the simultaneous assessment of these components may provide yet more useful information. Currently such multi-modal approaches are limited by the cost or technical demands that limit the number of individuals that can be studied, and thus limit the power and generalizability of the research. However, given that the price of nucleic acid sequencing has steadily decreased since the discovery of DNA, it is likely that the limits due to cost will eventually be lifted. Continued advances in the field of bioinformatics should also improve the feasibility of systems vaccinology approaches. Used in tandem with classical methods to confirm new discoveries, systems vaccinology has clear potential to support the development and delivery of safer and more effective vaccines.

References

Agnew-Crumpton, R,, Vaz, P.K., Devlin, J.M., O'Rourke, D,, Blacker-Smith, H.P., Konsak-Ilievski, B., Hartley, C.A., and Noormohammadi, A.H. (2016). Spread of the newly emerging infectious laryngotracheitis viruses in Australia. Infect Genet Evol. *43*, 67-73. DOI: 10.1016/j.meegid.2016.05.023.

Blanchard, M.T., Chen, C.I., Anderson, M., Hall, M.R., Barthold, S.W., and Stott, J.L. (2010). Vet Microbiol *144,* 177-182. DOI: 10.1016/j.vetmic.2010.01.002.

Blum, J.S., Wearsch, P.A., and Cresswell, P. (2013). Pathways of antigen processing. Annu Rev Immunol *31,* 443-473.

Borrow, R., Carlone, G.M., Rosenstein, N., Blake, M., Feavers, I., Martin, D., Zollinger, W., Robbins, J., Aaberge, I., Granoff, D.M., Miller, E., Plikaytis, B., van Alphen, L., Poolman, J., Rappuoli, R., Danzig, L., Hackell, J., Danve, B., Caulfield, M., Lambert, S., and Stephens, D. (2006). Neisseria meningitidis group B correlates of protection and assay standardization--international meeting report

Emory University, Atlanta, Georgia, United States, 16-17 March 2005. Vaccine *24,* 5093-5107.

Boylston, A. (2013). The origins of vaccination: myths and reality. J R Soc Med *106,* 351-354. DOI: 10.1177/0141076813499292.

Burton, D.R. (2017). What Are the Most Powerful Immunogen Design Vaccine Strategies? Reverse Vaccinology 2.0 Shows Great Promise. Cold Spring Harb Perspect Biol *9,* pii: a030262. DOI: 10.1101/cshperspect.a030262.

Capecchi, B., Serruto, D., Adu-Bobie, J., Rappuoli, R., and Pizza, M. (2004). The genome revolution in vaccine research. Curr Issues Mol Biol *6,* 17-27.

Chen, N., Trible, B.R., Kerrigan, M.A., Tian, K., Rowland, R.R.R. (2016). ORF5 of porcine reproductive and respiratory syndrome virus (PRRSV) is a target of diversifying selection as infection progresses from acute infection to virus rebound. Infect Genet Evol *40,* 167-175.

Carrasco Pro, S., Zimic, M., and Nielsen, M. (2014). Improved pan-specific MHC class I peptide-binding predictions using a novel representation of the MHC-binding cleft environment. Tissue Antigens *83*, 94-100. DOI: 10.1111/tan.12292.

Contreras, M., Villar, M., Artigas-Jerónimo, S., Kornieieva, L., Mytrofanov, S., and de la Fuente, J. (2018). A reverse vaccinology approach to the identification and characterization of Ctenocephalides felis candidate protective antigens for the control of cat flea infestations. Parasit Vectors *11,* 43. DOI: 10.1186/s13071-018-2618-x.

Correia, B.E., Bates, J.T., Loomis, R.J., Baneyx, G., Carrico, C., Jardine, J.G., Rupert, P., Correnti, C., Kalyuzhniy, O., Vittal, V., Connell, M.J., Stevens, E., Schroeter, A., Chen, M., Macpherson, S., Serra, A.M., Adachi, Y., Holmes, M.A., Li, Y., Klevit, R.E., Graham, B.S., Wyatt, R.T., Baker, D., Strong, R.K., Crowe, J.E. Jr., Johnson, P.R., and Schief, W.R. (2014). Proof of principle for epitope-focused vaccine design. Nature *507,* 201-2016. DOI: 10.1038/nature12966.

Crank, M.C., Ruckwardt, T.J., Chen, M., Morabito, K.M., Phung, E., Costner, P.J., Holman, L.A., Hickman, S.P., Berkowitz, N.M., Gordon, I.J., Yamshchikov, G.V., Gaudinski, M.R., Kumar, A., Chang, L.A., Moin, S.M., Hill, J.P., DiPiazza, A.T., Schwartz, R.M., Kueltzo, L., Cooper, J.W., Chen, P., Stein, J.A., Carlton, K., Gall, J.G., Nason, M.C., Kwong, P.D., Chen, G.L., Mascola, J.R., McLellan, J.S., Ledgerwood, J.E., Graham, B.S., and VRC 317 Study Team. (2019). A proof of concept for structure-based vaccine design targeting RSV in humans. Science. *365,* 505-509. DOI: 10.1126/science.aav9033.

Dalsass, M., Brozzi, A., Medini, D., and Rappuoli, R. (2019). Comparison of Open-Source Reverse Vaccinology Programs for Bacterial Vaccine Antigen Discovery. Front Immunol *14,* 113. DOI: 10.3389/fimmu.2019.00113.

Dash, P., Fiore-Gartland, A.J., Hertz, T., Wang, G.C., Sharma, S., Souquette, A., Crawford, J.C., Clemens, E.B., Nguyen, T.H.O., Kedzierska, K., La Gruta, N.L., Bradley, P., and Thomas, P.G. (2017). Quantifiable predictive features define epitope-specific T cell receptor repertoires. Nature. *547,* 89-93. DOI: 10.1038/nature22383.

Davis, M.M., and Boyd, S.D. (2019). Recent progress in the analysis of αβT cell and B cell receptor repertoires. Curr Opin Immunol *59,* 109-114.

DeKosky, B.J., Ippolito, G.C., Deschner, R.P., Lavinder, J.J., Wine, Y., Rawlings, B.M., Varadarajan, N., Giesecke, C., Dörner, T., Andrews, S.F., Wilson, P.C., Hunicke-Smith, S.P., Willson, C.G., Ellington, A.D., and Georgiou, G. (2013). High-throughput sequencing of the paired human immunoglobulin heavy and light chain repertoire. Nat Biotechnol *31*, 166-169.

DeLisi, C., and Berzofsky, J.A. (1985). T-cell antigenic sites tend to be amphipathic structures. Proc Natl Acad Sci U.S.A. *82,* 7048-7052.

Demasius, W., Weikard, R., Hadlich, F., Müller, K.E., and Kühn, C. (2013). Monitoring the immune response to vaccination with an inactivated vaccine associated to bovine neonatal pancytopenia by deep sequencing transcriptome analysis in cattle. Vet Res *44,* 93. DOI: 10.1186/1297-9716-44-93.

Dhanda, S.K., Usmani, S.S., Agrawal, P., Nagpal, G., Gautam, A., and Raghava, G.P.S. (2017). Novel in silico tools for designing peptide-based subunit vaccines and immunotherapeutics. Brief Bioinform *18*, 467-478. DOI: 10.1093/bib/bbw025.

Diaz-San Segundo, F., Medina, G.N., Stenfeldt, C., Arzt, J., and de Los Santos, T. (2017). Food-and-mouth disease vaccines. Vet Microbiol *206,* 102-112. DOI: 10.1016/j.vetmic.2016.12.018.

Ferdous, S., Kelm, S., Baker, T.S., Shi, J., and Martin, A.C.R. (2019). B-cell epitopes: Discontinuity and conformational analysis. Mol Immunol *114,* 643-650. DOI: 10.1016/j.molimm.2019.09.014.

Fisher, C.R., and Schnell, M.J. (2018). New developments in rabies vaccination. Rev Sci Technology *37,* 657-672. DOI: 10.20506/rst.37.2.2831.

Gilman, M.S., Castellanos, C.A., Chen, M., Ngwuta, J.O., Goodwin, E., Moin, S.M., Mas, V., Melero, J.A., Wright, P.F., Graham, B.S., McLellan, J.S., and Walker, L.M. (2016). Rapid profiling of RSV antibody repertoires from the memory B cells of naturally infected adult donors. Sci Immunol *1,* pii: eaaj1879. doi: 10.1126/sciimmunol.aaj1879.

Giuliani, M.M., Adu-Bobie, J., Comanducci, M., Aricò, B., Savino, S., Santini, L., Brunelli, B., Bambini, S., Biolchi, A., Capecchi, B., Cartocci, E., Ciucchi, L., Di Marcello, F., Ferlicca, F., Galli, B., Luzzi, E., Masignani, V., Serruto, D., Veggi, D., Contorni, M., Morandi, M., Bartalesi, A., Cinotti, V., Mannucci, D., Titta, F., Ovidi, E., Welsch, J.A., Granoff, D., Rappuoli, RI, and Pizza, M. (2006). A universal vaccine for serogroup B meningococcus. Proc Natl Acad Sci USA *103,* 10834-10839.

Gonzalez-Dias, P., Lee, E.K., Sorgi, S., de Lima, D.S., Urbanski, A.H., Silveira, E.L., and Nakaya, H,I. (2020). Methods for predicting vaccine immunogenicity and reactogenicity. Hum Vaccin Immunother *16,* 269-276. DOI: 10.1080/21645515.2019.1697110

Greenbaum, J.A., Andersen, P.H., Blythe, M., Bui, H.H., Cachau, R.E., Crowe, J., Davies, M., Kolaskar, A.S., Lund, O., Morrison, S., Mumey, B., Ofran, Y., Pellequer, J.L., Pinilla, C., Ponomarenko, J.V., Raghava, G.P., van Regenmortel, M.H., Roggen, E.L., Sette, A., Schlessinger, A., Sollner, J., Zand, M., and Peters, B. (2007). Towards a consensus on datasets and evaluation metrics for developing B-cell epitope prediction tools. J Mol Recognit *20,* 75-82.

Greenwood B. (2014). The contribution of vaccination to global health: past, present and future. Philos Trans R Soc Lond B Biol Sci *369,* 20130433. DOI: 10.1098/rstb.2013.0433.

Gupta, S., Ansari, H.R., Gautam, A., Open Source Drug Discovery Consortium, and Raghava, G.P. (2013). Identification of B-cell epitopes in an antigen for inducing specific class of antibodies. Biol Direct *8,* 27. DOI: 10.1186/1745-6150-8-27.

Gutiérrez, A.H., Martin, W.D., Bailey-Kellogg, C., Terry, F., Moise, L., and DeGroot, A.S. (2015). Development and validation of an epitope prediction tool for swine (PigMatrix) based on the pocket profile method. BMC Bioinformatics *16,* 290. DOI: 10.1186/s12859-015-0724-8.

Hagen, T., and Pulendran, B. (2019). Will systems biology deliver its promise and contribute to the development of new or improved vaccines? Cold Spring Harb Perspect Biol *10,* DOI:10.1101/cshperspect.a028894.

Haining, W.N. (2014). Strength in numbers: comparing vaccine signatures the modular way. Nat Immunol *15,* 139-141.

Hansen, A.M., Rasmussen, M., Svitek, N., Harndahl, M., Golde, W.T., Barlow, J., Vishvanath, N., Buus, S., and Nielsen, M. (2014). Immunogenetics *66*, 705-718. DOI: 10.1007/s00251-014-0802-5.

Hassan, A., Naz, A., Obaid A., Paracha, R.Z., Naz, K., Awan, F.M., Muhmmad, S.A., Janjua, H.A., Ahmad, J., and Ali, A. (2016). Pangenome and immuno-proteomics analysis of Acinetobacter baumannii strains revealed the core peptide vaccine targets. BMC Genomics *17,* 732. DOI: 10.1186/s12864-016-2951-4.

Hewitt, J.S., Karuppannan, A.K., Tan, S., Gauger, P., Halbur, P.G., Gerber, P.F., DeGroot, A.S., Moise, L., and Opriessnig, T. (2019). A prime-boost concept using a T-cell epitope-driven DNA vaccine followed by a whole virus vaccine effectively protected pigs in the pandemic H1N1 pig challenge model. Vaccine *37,* 4302-4309. DOI: 10.1016/j.vaccine.2019.06.044.

HIPC-CHI Signatures Project Team and HIPC-I Consortium. (2017). Multicohort analysis reveals baseline transcriptional predictors of influenza vaccination responses. Sci Immunol *2,* DOI:10.1126/sciimmunol.aal4656.

Hoof, K., Peters, B., Sidney, J., Pedersen, L.E., Sette, A., Lund, O., Buus, S., and Nielsen, M. (2009). NetMHCpan, a method for MHC class I binding prediction beyond humans. Immunogenetics *61*, 1-13. DOI: 10.1007/s00251-008-0341-z.

Jardetzky, T.S., Brown, J.H., Gorga, J.C., Stern, L.J., Urban, R.G., Strominger, J.L., and Wiley, D.C. (1996). Crystallographic analysis of endogenous peptides associated with HLA-DR1 suggests a common, polyproline II-like conformation for bound peptides. Proc Natl Acad Sci U.S.A. *23,* 734-738.

Jespersen, M.C., Peters, B., Nielsen, M., and Marcatili, P. (2017). BepiPred-2.0: improving sequence-based B-cell epitope prediction using conformational epitopes. Nucleic Acids Res *45,* W24-W29. DOI: 10.1093/nar/gkx346.

Jespersen, M.C., Mahajan, S., Peters, B., Nielsen, M., and Marcatili, P. (2019). Antibody Specific B-Cell Epitope Predictions: Leveraging Information From Antibody-Antigen Protein Complexes. Front Immunol *10,* 298. DOI: 10.3389/fimmu. 2019.00298.

Jurtz, V., Paul, S., Andreatta, M., Marcatilli, P., Peters, B., and Nielsen, M. (2017). NetMHCpan-4.0: improved peptide-MHC class I interaction predictions integrating eluted ligand and peptide binding affinity data. J Immunol *199*, 3360-3368. DOI: 10.4049/jimmunol.1700893.

Kapikian, A.Z., Mitchell, R.H., Chanock, R.M., Shvedoff, R.A., and Stewart, C.E. (1969). An epidemiologic study of altered clinical reactivity to respiratory syncytial

(RS) virus infection in children previously vaccinated with an inactivated RS virus vaccine. Am J Epidemiol *89,* 405-421.

Kringelum, J.V., Lundegaard, C., Lund, O., and Nielsen, M. (2012). Reliable B cell epitope predictions: impacts of method development and improved benchmarking. PLoS Comput Biol *8,* e1002829. DOI: 10.1371/journal.pcbi.1002829.

Kringelum, J.V., Nielsen, M., Padkjær, S.B., and Lund, O. (2013). Structural analysis of B-cell epitopes in antibody:protein complexes. Mol Immunol *53,* 24-34. DOI: 10.1016/j.molimm.2012.06.001.

Li, S., Rouphael, N., Duraisingham, S., Romero-Steiner, S., Presnell, S., Davis, C., Schmidt, D.S., Johnson, S.E., Milton, A., Rajam, G., Kasturi, S., Carlone, G.M., Quinn, C., Chaussabel, D., Palucka, A.K., Mulligan, M.J., Ahmed, R., Stephens, D.S., Nakaya, H.I., and Pulendran, B. (2014). Molecular signatures of antibody responses derived from a systems biology study of five human vaccines. Nat Immunol *15,* 195-204. DOI: 10.1038/ni.2789.

Li, J.N., Zhao, Y.T., Cao, S.L., Wang, H., and Zhang, J.J. (2020). Integrated transcriptomic and proteomic analyses of grass carp intestines after vaccination with a double-targeted DNA vaccine of Vibrio mimicus. Fish Shellfish Immunol *98,* 641-652. DOI: 10.1016/j.fsi.2019.10.045.

Lund, O., Nielsen, M., Kesmir, C., Petersen, A.G., Lundegaard, C., Worning, P., Sylvester-Hvid, C., Lamberth, K., Røder, G., Justesen, S., Buus, S., and Brunak, S. (2004). Definition of supertypes for HLA molecules using clustering of specificity matrices. Immunogenetics *55*, 797-810. Doi: 10.1007/s00251-004-0647-4.

Lundegaard, C., Hoof, I., Lund, O., and Nielsen, M. (2010). State of the art and challenges in sequence based T-cell epitope prediction. Immunome Res *6,* Suppl 2:S3. DOI: 10.1186/1745-7580-6-S2-S3.

Luo, J., Carrillo, J.A., Menendez, K.R., Tablante, N.L., and Song, J. (2014). Transcriptome analysis reveals an activation of major histocompatibility complex 1 and 2 pathways in chicken trachea immunized with infectious laryngotracheitis virus vaccine. Poult Sci *93,* 848-855. DOI: 10.3382/ps.2013-03624.

Magro, M., Mas, V., Chappell, K., Vázquez, M., Cano, O., Luque, D., Terrón, M.C., Melero, J.A., and Palomo, C. (2012). Neutralizing antibodies against the preactive form of respiratory syncytial virus fusion protein offer unique possibilities for clinical intervention. Proc Natl Acad Sci U.S.A. *109,* 3089-3094. DOI: 10.1073/pnas.1115941109.

Madden, D.R., (1995) The three-dimensional structure of peptide-MHC complexes. Annu Rev Immunol *13*, 587-622.

Margalit, H., Spouge, J.L., Cornette, J.L., Cease, K.B., Delisi, C., and Berzofsky, J.A.. (1987). Prediction of immunodominant helper T cell antigenic sites from the primary sequence. J Immunol *138,* 2213-2229.

Matthijs, A.M.F., Auray, G., Jakob, V., García-Nicolás, O., Braun, R.O., Keller, I., Bruggman, R., Devriendt, B., Boyen, F., Guzman, C.A., Michiels, A., Haesebrouck, F., Collin, N., Barnier-Quer, C., Maes, D., and Summerfield, A. (2019). Systems Immunology Characterization of Novel Vaccine Formulations for Mycoplasma hyopneumoniae Bacterins. Front Immunol *10,* 1087. DOI: 10.3389/fimmu. 2019.01087.

Mei, S., Li, F., Leier, A., Marquez-Lago, T.T., Giam, K., Croft, N.P., Akutsu, T., Smith, A.I., Li, J., Rossjohn, J., Purcell, A.W., and Song, J. (2019). A comprehensive review and performance evaluation of bioinformatics tools for HLA class I peptide-binding prediction. Brief Bioinform June 14 bbz051. DOI: 10.1093/bib/bbz051.

Mitsunaga, E.M., and Snyder, M.P. (2020). Deep Characterization of the Human Antibody Response to Natural Infection Using Longitudinal Immune Repertoire Sequencing. Mol Cell Proteomics *19,* 278-293. DOI: 10.1074/mcp.RA119.001633.

Moxon, R., Reche, P.A., and Rappuoli, R. (2019). Reverse vaccinology. Front Immunol *3,* 2776. DOI: 10.3389/fimmu.2019.02776.

Nakaya, H.I., Wrammert, J., Lee, E.K., Racioppi, L., Marie-Kunze, S., Haining, W.N., Means, A.R., Kasturi, S.P., Khan, N., Li, G.M., McCausland, M., Kanchan, V., Kokko, K.E., Li, S., Elbein, R., Mehta, A.K., Aderem, A., Subbarao, K., Ahmed, R., and Pulendran, B. (2011). Systems biology of vaccination for seasonal influenza in humans. Nat Immunol *12,* 786-795. DOI: 10.1038/ni.2067.

Nielsen, M., Connelley, T., and Ternette, N. (2018). Improved prediction of bovine leucocyte antigens (BoLA) presented ligands by use of mass-spectrometry-determined ligand and in vitro binding data. J Proteome Res *17*, 559-567. Doi: 10.1021/acs.jproteome.7b00675.

Novotný, J., Handschumacher, M., Haber, E., Bruccoleri, R.E., Carlson, W.B., Fanning, D.W., Smith, J.A., and Rose, G.D. (1986). Antigenic determinants in proteins coincide with surface regions accessible to large probes (antibody domains). Proc Natl Acad Sci U.S.A. *83,* 226-230.

Oh, J.Z., Ravindran, R., Chassaing, B., Carvalho, F.A., Maddur, M.S., Bower, M., Hakimpour, P., Gill, K.P., Nakaya, H.I., Yarovinsky, F., Sartor, R.B., Gewirtz, A.T., and Pulendran, B. (2014). TLR5-mediated sensing of gut microbiota is necessary for antibody responses to seasonal influenza vaccination. Immunity *41,* 478-492. DOI: 10.1016/j.immuni.2014.08.009.

Painter, C.A., and Stern, L.J. (2012). Conformational variation in structures of classical and non-classical MHCII proteins and functional implications. Immunol Rev *250,* 144-157.

Pandya, M., Rasmussen, M., Hansen, A., Nielsen, M., Buus, S., Golde, W., and Barlow, J. (2015). A modern approach for epitope prediction: identification of food-and-mouth disease virus peptides binding bovine leukocyte antigen (BoLA) class I molecules. Immunogenetics *67*, 691-703. Doi: 10.1007/s00251-015-0877-7.

Pedersen, L.E., Harndahl, M., Nielsen, M., Patch, J.R., Jungersen, G., Buus, S., and Golde, W.T. (2013). Identification of peptides from foot-and-mouth disease virus structural proteins bound by class I swine leukocyte antigen (SLA) alleles, SLA-1*0401 and SLA-2*0401. Anim Genet *44,* 251-258. DOI: 10.1111/j.1365-2052.2012.02400.x.

Pizza, M., Scarlato, VI, Masignani, V., Giuliani, M.M., Aricò, B., Comanducci, M., Jennings, G.T., Baldi, L., Bartolini, E., Capecchi, B., Galeotti, C.L., Luzzi, E., Manetti, R., Marchetti, E., Mora, M., Nuti, S., Ratti, G., Santini, L., Savino, S., Scarselli, M., Storni, E., Zuo, P., Broeker, M., Hundt, E., Knapp, B., Blair, E., Mason, T., Tettelin, H., Hood, D.W., Jeffries, A.C., Saunders, N.J., Granoff, D.M., Venter, J.C., Moxon, E.R., Grandi, G., and Rappuoli, R. (2000). Identification of vaccine candidates against serogroup B meningococcus by whole-genome sequencing. Science *10,* 1816-1820.

Pretorius, A., Faber, F.E., and van Kleef, M. (2016). Immune gene expression profiling of PBMC isolated from horses vaccinated with attenuated African horsesickness virus serotype 4. Immunobiology *221,* 236-244. DOI: 10.1016/j.imbio.2015.09.002

Pulendran, B., Li, S., and Nakaya, H.I. (2010). Systems vaccinoilogy. Imunity *33,* 516-529. DOI: 10.1016/j.immuni.2010.10.006.

Querec, T.D., Akondy, R.S., Lee, E.K., Cao, W., Nakaya, H.I., Teuwen, D., Pirani, A., Gernert, K., Deng, J., Marzolf, B., Kennedy, K., Wu, H., Bennouna, S., Oluoch, H., Miller, J., Vencio, R.Z., Mulligan, M., Aderem, A., Ahmed, R., and Pulendran, B1. (2009) Systems biology approach predicts immunogenicity of the yellow fever vaccine in humans. Nat Immunol *10,* 116-125. DOI: 10.1038/ni.1688.

Rahman, Kh.S., Chowdhury, E.U., Sachse, K, and Kaltenboeck, B. (2016). Inadequate Reference Datasets Biased toward Short Non-epitopes Confound B-cell Epitope Prediction. J Biol Chem *291,* 14585-14599. DOI: 10.1074/jbc.M116.729020.

Rammensee, H.G., Friede, T., and Stevanoviic, S. (1995). MHC ligands and peptide motifs: first listing. Immunogenetics *41,* 178-228.

Rappuoli, R. (2000). Reverse vaccinology. Curr Opin Microbiol *3,* 445-450.

Rappuoli, R. (2007). Bridging the knowledge gaps in vaccine design. Nat Biotechnol. *25,* 1361-1366.

Rappuoli, R., Bottomley, M.J., D'Oro, U., Finco, O., and De Gregorio, E. (2016). Reverse vaccinology 2.0: Human immunology instructs vaccine antigen design. J Exp Med *213,* 469-481. DOI: 10.1084/jem.20151960.

Reyes, V.E., Chin, L.T., and Humphreys, R.E. (1988). Selection of class I MHC-restricted peptides with the strip-of-helix hydrophobicity algorithm. Mol Immunol *25,* 867-871.

Robinson, J., Guethlein, L.A., Cereb, N., Yang, S.Y., Norman, P.J., Marsh, S.G.E., and Parham, P. (2017). Distinguishing functional polymorphism from random variation in the sequences of >10,000 HLA-A, -B and -C alleles. PLoS Genet *13,* e1006862. DOI: 10.1371/journal.pgen.1006862.

Rock, K.L., Reits, E., and Neefjes, J. (2016). Present Yourself! By MHC Class I and MHC Class II Molecules. Trends Immunol *37,* 724-737. DOI: 10.1016/j.it.2016.08.010.

Saada, R., Weinberger, M., Shahaf, G., and Mehr, R. (2007). Models for antigen receptor gene rearrangement: CDR3 length. Immunol Cell Biol *85,* 323-332.

Sant, S., Grzelak, L., Wang, Z., Pizzolla, A., Koutsakos, M., Crowe, J., Loudovaris, T., Mannering, S.I., Westall, G.P., Wakim, L.M., Rossjohn, J., Gras, S., Richards, M., Xu, J., Thomas, P.G., Loh, L., Nguyen, T.H.O., and Kedzierska, K. (2018). Single-Cell Approach to Influenza-Specific CD8+ T Cell Receptor Repertoires Across Different Age Groups, Tissues, and Following Influenza Virus Infection. Front Immunol *9,* 1453. DOI: 10.3389/fimmu.2018.01453.

Saravanan, V. and Gautham, N. (2018). BCIgEPRED-a dual-layer approach for predicting linear IgE epitopes. Mol Biol (Mosk) *52,* 333-343. DOI: 10.7868/S0026898418020180.

Sela-Culang, I., Ofran, Y., and Peters, B. (2015). Antibody specific epitope prediction-emergence of a new paradigm. Curr Opin Virol *11,* 98-102. DOI: 10.1016/j.coviro. 2015.03.012.

Serruto, D., Bottomley, M.J., Ram, S., Giuliani, M.M., and Rappuoli, R. (2012). The new multicomponent vaccine against meningococcal serogroup B, 4CMenB: immunological, functional and structural characterization of the antigens. Vaccine *30,* 30. DOI: 10.1016/j.vaccine.2012.01.033.

Stave, J.W. and Lindpainter, K. (2013). Antibody and antigen contact residues define epitope and paratope size and structure. J Immunol *191,* 1428-1435. DOI: 10.4049/jimmunol.1203198.

Stern, L.J., Brown, J.H., Jardetzky, T.S., Gorga, J.C., Urban, R.G., Strominger, J.L., and Wiley, D.C. (1994). Crystal structure of the human class II MHC protein HLA-DR1 complexed with an influenza virus peptide. Nature *368,* 215-221.

Stille, C.J., Thomas, L.J., Reyes, V.E., and Humphreys, R.E. (1987). Hydrophobic strip-of-helix algorithm for selection of T cell-presented peptides. Mol Immunol *24,* 1021-1027.

Sundling, C., Martinez, P., Soldemo, M., Spångberg, M., Bengtsson, K.L., Stertman, L., Forsell, M.N., and Karlsson Hedestam, G.B. (2013). Immunization of macaques with soluble HIV type 1 and influenza virus envelope glycoproteins results in a similarly rapid contraction of peripheral B-cell responses after boosting. J Infect Dis *207,* 426-431. DOI: 10.1093/infdis/jis696.

Svitek, N., Hansen, A.M., Steinaa, L., Saya, R., Awino, E., Nielsen, M., Buus, S., and Nene, V. (2014). Use of "one-pot, mix-and-read" peptide-MHC class I tetramers and predictive algorithms to improve detection of cytotoxic T lymphocyte response in cattle. Vet Res *45,* 50. DOI: 10.1186/1297-9716-45-50.

Terasaki, P.I. (2007). A brief history of HLA. Immunol Res *38,* 139-148. DOI: 10.1007/ s12026-007-0020-4

Tettelin, H,, Saunders, N.J., Heidelberg, J., Jeffries, A.C., Nelson, K.E., Eisen, J.A., Ketchum, K.A., Hood, D.W., Peden, J.F., Dodson, R.J., Nelson, W.C., Gwinn, M.L., DeBoy, R., Peterson, J.D., Hickey, E.K., Haft, D.H., Salzberg, S.L., White, O., Fleischmann, R.D., Dougherty, B.A., Mason, T., Ciecko, A., Parksey, D.S., Blair, E., Cittone, H., Clark, E.B., Cotton, M.D., Utterback, T.R., Khouri, H., Qin, H., Vamathevan, J., Gill, J., Scarlato, V., Masignani, V., Pizza, M., Grandi, G., Sun, L., Smith, H.O., Fraser, C.M., Moxon, E.R., Rappuoli, R., and Venter, J.C. (2000).

Complete genome sequence of Neisseria meningitidis serogroup B strain MC58. Science *287,* 1809-1815.

Tomazic, M.L., Rodriguez, A.E., Lombardelli, J., Poklepovich, T., Garro, C., Galarza, R., Tiranti, K., Florin-Christensen, M., and Schnittger, L. (2018). Identification of novel vaccine candidates against cryptosporidiosis of neonatal bovines by reverse vaccinology. Vet Parasitol *264,* 74-78. DOI: 10.1016/j.vetpar.2018.11.007.

Vahey, M.T., Wang, Z., Kester, K.E., Cummings, J., Heppner, D.G. Jr, Nau, M.E., Ofori-Anyinam, O., Cohen, J., Coche, T., Ballou, W.R., and Ockenhouse, C.F. (2010). Expression of genes associated with immunoproteasome processing of major histocompatibility complex peptides is indicative of protection with adjuvanted RTS,S malaria vaccine. J Infect Dis *201,* 580-589. DOI: 10.1086/650310.

Van Regenmortel, M.H. (2014). Specificity, polyspecificity, and heterospecificity of antibody-antigen recognition. J Mol Recognit *27,* 627-639. DOI: 10.1002/jmr.2394.

Waickman, A.T., Victor, K., Li, T., Hatch, K., Rutvisuttinunt, W., Medin, C., Gabriel, B., Jarman, R.G., Friberg, H., and Currier, J.R. (2019). Dissecting the heterogeneity of DENV vaccine-elicited cellular immunity using single-cell RNA sequencing and metabolic profiling. Nat Commun *10,* 3666. DOI: 10.1038/s41467-019-11634-7.

Welly, B.T., Miller, M.R., Stott, J.L., Blanchard, M.T., Islas-Trejo, A.D., O'Rourke, S.M., Young, A.E., Medrano, J.F., and Van Eenennaam, A.L. (2017). Genome Report: Identification and Validation of Antigenic Proteins from Pajaroellobacter abortibovis Using De Novo Genome Sequence Assembly and Reverse Vaccinology. G3. *7,* 321-331. DOI: 10.1534/g3.116.036673.

Chapter 2

Cancer Vaccines

Philip J. Bergman*

Oncologist and VCA Director of Clinical Studies, Katonah Bedford Veterinary Center, Bedford Hills, NY, USA and Adjunct Associate member, Memorial Sloan-Kettering Cancer Center, New York, NY, USA

*Corresponding author: philip.bergman@vca.com

DOI: https://doi.org/10.21775/9781913652593.02

Abstract

The enhanced understanding of immunology experienced over the last 4-5 decades afforded through the tools of molecular biology has recently translated into cancer immunotherapy becoming one of the most exciting and rapidly expanding fields. Human cancer immunotherapy is now recognized as one of the pillars of treatment alongside surgery, radiation and chemotherapy. The field of veterinary cancer immunotherapy has also rapidly advanced in the last decade with a handful of commercially available products (one of which is a xenogeneic DNA cancer vaccine) and a plethora of investigational cancer immunotherapies which will likely expand our veterinary oncology treatment toolkit over time.

Introduction

The term "immunity" is derived from the Latin word immunitas, which refers to the legal protection afforded to Roman senators holding office. While the immune system is normally thought of as providing protection against infectious disease, the immune system's ability to recognize and eliminate cancer is the fundamental rationale for the immunotherapy of cancer. Multiple lines of evidence support a role for the immune system in managing cancer, including: 1) spontaneous remissions in cancer patients without treatment, 2) the presence of tumor-specific cytotoxic T-cells within tumor or draining lymph nodes, 3) the presence of monocytic, lymphocytic and plasmacytic cellular infiltrates in tumors, 4) the increased incidence of some types of cancer in immunosuppressed patients and 5) documentation of cancer remissions with the use of immunomodulators (Bergman 2001, Baxevanis et al 2009). With the tools of molecular biology and a greater understanding of mechanisms to harness the immune system, effective tumor immunotherapy is now a reality. This new class of therapeutics offers a more targeted and therefore precise approach to the treatment of cancer. Cancer immunotherapy is now recognized as one of the pillars of treatment alongside surgery, radiation and chemotherapy.

Tumor Immunology

Cellular Components

The immune system is generally divided into **two primary components**: the *innate immune response*, and the highly specific, but more slowly developing *adaptive or acquired immune response.* Innate immunity is rapidly acting but typically not very specific and includes physico-chemical barriers (e.g. skin and mucosa), blood proteins such as complement, phagocytic cells (macrophages, neutrophils, dendritic cells (DCs), and natural killer (NK) cells) and cytokines which coordinate and regulate the cells involved in innate immunity. Adaptive immunity is thought of as the acquired arm of immunity which allows for exquisite specificity, an ability to remember the previous existence of the pathogen (i.e. “memory”), differentiate self from non-self, and importantly the ability to respond more vigorously upon repeat exposure to the pathogen. Adaptive immunity consists of T and B lymphocytes. The T cells are further divided into CD8 (cluster of differentiation) and MHC (major histocompatibility complex) Class I cytotoxic helper T cells (CD4 and MHC class II), natural killer (NK) cells and regulatory T cells. B lymphocytes produce antibodies (humoral system) which may activate complement, enhance phagocytosis of opsonized target cells and induce antibody dependent cellular cytotoxicity (ADCC). B-cell responses to tumors are thought by many investigators to be less important than the development of T-cell mediated immunity, but there is little evidence to fully support this notion (Reilly et al 2001). The innate and adaptive arms of immunity are not mutually exclusive; they are linked by: 1) the innate response’s ability to stimulate and influence the nature of the adaptive response, and 2) the sharing of effector mechanisms between innate and adaptive immune responses.

Immune responses can be further separated by whether they are induced by exposure to a foreign antigen (an "active" response) or if they are transferred through serum or lymphocytes from an immunized individual (a "passive" response). While both approaches have the ability to be extremely specific for an antigen of interest, one important difference is the inability of passive approaches to generally confer memory. The principal components of the active/adaptive immune system are lymphocytes, antigen-presenting cells and effector cells. Furthermore, responses can be sub-divided by whether they are specific for a certain antigen, or a non-specific response whereby immunity is attempted to be conferred by upregulating the immune system without a specific target. These definitions are helpful as they allow methodologies to be more completely characterized, such as active-specific, passive-nonspecific, etc.

Immune Surveillance

The idea that the immune system may actively prevent the development of neoplasia is termed "cancer immunosurveillance." Sound scientific evidence supports some aspects of this hypothesis (Smyth et al 2001, Wallace et al 2005, Itoh et al 2009, Schmiedt et al 2009) including: 1) IFN-γ protects mice against the growth of tumors, 2) mice lacking IFN-γ receptor were more sensitive to chemically-induced sarcomas than normal mice and were more likely to spontaneously develop tumors, 3) mice lacking major components of the adaptive immune response (T and B cells) have a high rate of spontaneous tumors, and 4) mice that lack IFN-γ and B/T cells develop tumors, especially at a young age.

Immune Evasion by Tumors

There are significant barriers to the generation of effective antitumor immunity by the host. Many tumors evade surveillance mechanisms and grow in immunocompetent hosts as easily illustrated by the overwhelming numbers of people and animals succumbing to cancer. There are multiple ways in which tumors evade the immune response including: 1) Immunosuppressive cytokine production (e.g. TGF-β and IL-10)(Catchpole et al., 2002; Zagury and Gallo, 2004), 2) Impaired DC function via inactivation ("anergy") and/or poor DC maturation through changes in IL-6/IL-10/ VEGF/GM-CSF(Morse et al., 2002), 3) Induction of cells called Regulatory T cells (Treg), which were initially called suppressor T cells (CD4/CD25/CTLA-4/GITR/ Foxp3 positive cells which can suppress tumor-specific CD4/CD8+ T cells)

(Yamaguchi and Sakaguchi, 2006), 4) MHC I loss through structural defects, changes in B2-microglobulin synthesis, defects in transporter-associated antigen processing or actual MHC I gene loss (ie allelic or locus loss) and 5) MHC I antigen presentation loss through B7-1 attenuation (B7-1 is an important co-stimulatory molecule for CD28 mediated T-cell receptor and MHC engagement) when the MHC system in #4 remains intact.

Cancer Vaccines

Overview

The ultimate goal for a tumor immunotherapy with a specific target is elicitation of an anti-tumor immune response which results in clinical regression of a tumor and/or its metastases. There are numerous types of specific tumor immunotherapies in phase I-III trials across a wide range of tumor types. Responses to cancer vaccines and other cancer immunotherapies may take several months or more to appear due to the slower speed of induction of the adaptive arm of the immune system as outlined in Table 1. This has necessitated the development of an alternative and more immunotherapeutic-based response system for human studies and this is highly likely necessary in the future for veterinary studies (Seymour et al., 2017). The ideal cancer immunotherapy agent would be able to discriminate between cancer and normal cells (ie specificity), be potent enough to kill small or large numbers of tumor cells (ie sensitivity) and lastly be able to prevent recurrence of the tumor (ie durability).

The immune system detects tumors through specific tumor-associated antigens (TAAs) and/or abnormal disease-associated antigens (DAAs) that are potentially

Table 1. Comparison of chemotherapy and various anti-tumor immunotherapies

Treatment Type	Mechanism of Action	TAA or Target Dependent	Specificity	Sensitivity	Response Time	Durability of Response
Chemotherapy	Cytotoxicity	No	Poor	Variable	Hours - Days	Variable
Anti-Tumor Vaccine	Immune Response	Yes	Good	Good	Weeks - Months	Variable - Long
Monoclonal Antibodies	Immune Response	Yes	Good	Good	Weeks	Variable
Checkpoint Inhibitors	Immune Response	No	Low	Moderate	Weeks - Months	Long

recognized by both CTLs and antibodies (Bergman, 2007; Beatty and Finn, 2013; Regan et al., 2016). TAAs and/or DAAs may be common to a particular tumor type, be unique to an individual tumor or may arise from mutated gene products such as ras, p53, p21 and/or others. While unique TAAs may be more immunogenic than the other aforementioned shared tumor antigens, they are not practical targets because of their narrow specificity. Most shared tumor antigens are normal cellular antigens that are overexpressed in tumors. The first group to be identified was termed cancer testes antigens due to expression in normal testes, but they are also found in melanoma, and various other solid tumors such as the MAGE/BAGE gene family. This chapter will highlight those approaches which appear to hold particular promise in human clinical trials and many which have been tested to date in veterinary medicine.

Approaches

A number of investigations using recombinant cytokines, growth factors or hormones in various fashions for human and veterinary cancer patients have been reported to date. Many have investigated the in vitro and/or in vivo effects of the soluble cytokine (e.g. interferons, IL-2, IL-12, IL-15, etc. with or without suicide gene therapy), (Tateyama et al., 1995; Kruth, 1998; Whitley et al., 1995; Hampel et al., 2007; Finocchiaro and Glikin, 2008; Cutrera et al., 2008; Finocchiaro et al., 2008; Akhtar et al., 2004; Dickerson et al., 2002; Okano and Yamada, 2000; Jahnke et al., 2007; Dickerson et al., 2004; Finocchiaro et al., 2015; Finocchiaro et al., 2018) liposome encapsulation of the cytokine (e.g. liposomal IL-2),(Dow et al., 2005; Skubitz and Anderson, 2000; Khanna et al., 1997; Khanna et al., 1996; Kruth, 1998) or use a virus, cell, liposome-DNA complex, plasmid or other mechanism to expresses the cytokine (e.g. recombinant poxvirus expressing IL-2) (Jourdier et al., 2003; Siddiqui et al., 2006; Quintin-Colonna et al., 1996; Dow et al., 2005; Kamstock et al., 2006; Junco et al., 2008; Chou et al., 2009; Chuang et al., 2009; Finocchiaro and Glikin, 2012). Important to note, the European Committee for Medicinal Products for Veterinary Use (CVMP) adopted a positive opinion in March, 2013 for the veterinary product Oncept IL-2 (feline Pox virus expressing recombinant feline IL-2). This product also received conditional licensure from the USDA CVB (Center for Veterinary Biologics) in 2015. It is labelled for use in addition to surgery and radiation in cats with stage I fibrosarcomas without metastasis or lymph node involvement, to reduce the risk of relapse and increase the time to relapse.

A variety of approaches have been taken to date to focus the immune system on the aforementioned targets, including 1) Whole cell, tumor cell lysate and/or subunit vaccines (autologous, or made from a patient's own tumor tissue; allogeneic, or made from individuals within a species bearing the same type of cancer; or whole cell vaccines from γ-irradiated tumor cell lines with or without immunostimulatory cytokines)(Hogge et al., 1999; Alexander et al., 2006; Quintin-Colonna et al., 1996; U'Ren et al., 2007; Bird et al., 2008; Turek et al., 2007; Kuntsi-Vaattovaara et al., 2003; Milner et al., 2006; Marconato et al., 2013; Suckow, 2013; Epple et al., 2013; Andersen et al., 2013; Yannelli et al., 2016; Marconato et al., 2015; Marconato et al., 2014; Weir et al., 2014; Weir et al., 2018), 2) DNA vaccines which immunize with syngeneic and/or xenogeneic (different species than recipient) plasmid DNA designed to elicit antigen-specific humoral and cellular immunity(Kamstock et al., 2007; Yu et al., 2011; Impellizeri et al., 2012; Denies et al., 2016; Gabai et al., 2014; Riccardo et al., 2014; Gibson et al., 2015; Piras et al., 2017) (to be discussed in more detail later in this chapter), 3) Viral, virus-like nanoparticle and/or viral vector-based methodologies designed to be oncolytic and/or deliver genes encoding TAAs, telomerase and/or immunostimulatory cytokines,(von et al., 2008; Johnston et al., 2005; Thacker et al., 2009; Peruzzi et al., 2010; Gavazza et al., 2013; Autio et al., 2015; Impellizeri et al., 2018; Hoopes et al., 2018; Sanchez et al., 2018; Cejalvo et al., 2018; Hoopes et al., 2017) 4) DC or CD40-activated B cell vaccines (which are commonly loaded or transfected with TAAs, DNA or RNA from TAAs, or tumor lysates)(Gyorffy et al., 2005; Tamura et al., 2007; Tamura et al., 2008; Rodriguez-Lecompte et al., 2004; Kyte et al., 2006; Mason et al., 2008; Sorenmo et al., 2011; Bird et al., 2011; Hernandez-Granados et al., 2018) and adoptive cell transfer (the "transfer" of specific populations of immune effector cells in order to generate a more powerful and focused antitumor immune response; to be discussed in more detail later in this chapter with clinically relevant recent advances) (O'Connor et al., 2012).

One particularly exciting vaccine approach targets HER-2 with an attenuated listeria therapeutic vaccine or a DNA vaccine approach (Shahabi et al., 2011; Fazekas et al., 2016; Gibson et al., 2015). The listeria methodology was utilized by Mason et al in dogs with appendicular osteosarcoma after being treated with amputation and adjuvant carboplatin chemotherapy (Mason et al., 2016) The results from this Phase I study are particularly exciting as it translated into a median survival time approaching 3 years. It is currently unknown how much of the therapeutic efficacy is from the xenogeneic human HER-2 vs the listeria, but may be considered in the future with other HER-2 related histologies. This product from Aratana ("AT-014") received USDA-CVB conditional licensure in early 2018 and underwent an additional multi-center clinical safety study to potentially move the product to full licensure. Recent studies have documented Listeria colonization in a small number of treated dogs and the product is no longer commercially available (Musser et al., 2019)

This author has developed a xenogeneic DNA vaccine program for melanoma in collaboration with human investigators from Memorial Sloan-Kettering Cancer Center (Bergman et al., 2006; Bergman et al., 2003). Preclinical and clinical studies by our laboratory and others have shown that xenogeneic DNA vaccination with tyrosinase family members (e.g. Tyrosinase, GP100, GP75, others) can produce immune responses resulting in tumor rejection or protection and prolongation of survival while syngeneic vaccination with orthologous DNA does not induce immune responses. While tyrosinase may not appear to be a preferred target in amelanotic canine melanoma due to poor expression when assessed by immunohistochemistry (IHC), (Smedley et al., 2011) more appropriate/sensitive PCR-based studies and other IHC-based studies document significant tyrosinase overexpression in melanotic and amelanotic melanomas across species (Phillips et al., 2012; Cangul et al., 1999; Ramos-Vara et al., 2000; Ramos-Vara and Miller, 2011; de Vries et al., 2001; Gradilone et al., 2012). These studies provided the impetus for development of a xenogeneic tyrosinase (or similar melanosomal glycoproteins) DNA vaccine program in canine malignant melanoma (CMM). Cohorts of dogs received increasing doses of xenogeneic plasmid DNA encoding either human Tyrosinase (huTyr), murine GP75 (muGP75), murine tyrosinase (muTyr), muTyr ± HuGM-CSF (both administered as plasmid DNA) or muTyr "off-study" intramuscularly biweekly for a total of four vaccinations. We and collaborators have investigated the antibody and T-cell responses in dogs vaccinated with huTyr. Antigen-specific (huTyr) IFNγ T-cells were found along with 2- to 5-fold increases in circulating antibodies to huTyr which can cross react to canine Tyrosinase, suggesting the breaking of tolerance(Liao et al., 2006; Goubier et al., 2008). The clinical results with prolongation in survival have been reported previously(Bergman et al., 2006; Bergman et al., 2003). The results of these trials demonstrate that xenogeneic DNA vaccination in CMM: 1) is safe, 2) leads to the development of anti-tyrosinase antibodies and T cells, 3) is potentially therapeutic, and 4) is an attractive candidate for further evaluation in an adjuvant, minimal residual disease Phase II setting for CMM. Based on these studies a multi-institutional safety and efficacy trial for USDA licensure in dogs with locally controlled stage II/III oral melanoma was initiated in 2006 with granting of conditional licensure in 2007, which represented the first U.S. governmental regulatory agency approval of a vaccine to treat cancer across species. Results of this licensure trial documented a statistically significant improvement in survival for vaccinates vs controls and a full licensure for the HuTyr-based canine melanoma vaccine from USDA-CVB was received in December, 2009 (Oncept™, Merial, Inc.) (Grosenbaugh et al., 2011). Recently, other investigators have reported safety and feasibility results of a pilot study in dogs with melanoma given huTyr plasmid DNA (ie similar to Oncept™), but instead of the typical needle-free transdermal delivery, a novel tatoo-like "micro-seeding" delivery system was utilized and was well tolerated with no toxicity noted in the six dogs (Zuleger et al., 2017)

Kaser-Hotz and colleagues reported on concurrent use of Oncept™ and external beam radiation as many dogs with oral malignant melanoma may not be able to undergo surgery for local tumor control (Herzog et al., 2013). This pilot study determined that concurrent use was well tolerated with no unexpected toxicities. Ottnod and colleagues performed a single-site retrospective study on 30 dogs with stage II – III oral malignant melanoma (15 each with and without use of Oncept™). They determined that those dogs receiving Oncept™ did not achieve a greater progression-free survival, disease-free interval or median survival time than dogs that did not receive the vaccine (Ottnod et al., 2013). Contrary to the aforementioned prospective USDA 5-site licensure trial(Grosenbaugh et al., 2011), this study had < 35% of cases treated surgically with margins 1mm or more, suggesting a significant lack of local tumor control. Furthermore, contrary to the aforementioned prospective USDA 5-site licensure trial, the Ottnod et al study, similar to other non-controlled retrospective studies, either had a wide variety of other treatments utilized in both the non-vaccinate and vaccinate groups, had small numbers of patients investigated and/or the cause (in the context of local or distant disease) of death and/or progression of disease was not reported (Verganti et al., 2017; Treggiari et al., 2016; McLean and Lobetti, 2015; Boston et al., 2014). This author reported at the 2016 VCS meeting the outcomes of 320 dogs with malignant melanoma treated with Oncept™ across VCA oncology centers. The long term median outcomes noted in that study are extremely similar to those we reported from the USDA 5-site prospective licensure trial and compare highly favorably to outcomes reported with standardized therapies without Oncept™ . Not surprisingly, the smaller poorly-controlled retrospective studies do not appear to mirror the results seen in larger and more highly controlled studies.

Human clinical trials utilizing various xenogeneic melanosomal antigens as DNA (or peptide with adjuvant) vaccination began in 2005 and the preliminary results look favorable (Wolchok et al., 2007; Perales et al., 2008; Yuan et al., 2009). To further highlight xenogeneic DNA vaccination as a platform to target other possible antigens for other histologies, we have completed a Phase I trial of murine CD20 for dogs with B cell lymphoma (USDA-CVB conditionally licensed as Canine Lymphoma Vaccine from Merial Inc. and undergoing currently further efficacy studies in larger numbers of dogs) and we have also investigated the efficacy of local tumor control and use of xenogeneic DNA vaccination in dogs with digit malignant melanoma (Manley et al., 2011). These investigations led to the development of a canine digit melanoma staging scheme and found an improvement in survival compared to historical outcomes with digit amputation only. We also documented a decreased prognosis for dogs with advanced stage disease and/or increased time from digit amputation to the start of vaccination. Phillips and co-investigators have also reported the overexpression of tyrosinase in equine melanoma, determined the safety and optimal use of the needle-free delivery device into the pectoral region with Oncept™ and documented antigen-specific humoral responses after vaccination in all horses (Phillips et al., 2012; Phillips et al., 2011) Oncept also appears to be safe for use in cats (Sarbu et al., 2015)

A small subset of dogs with malignant melanoma have exon 11 KIT gene mutations, (Chu et al., 2012; Murakami et al., 2011) and therefore the more routine use of KIT testing by PCR of CMM and subsequent use of c-kit small molecule inhibitors (particularly in dogs with advanced stage disease and/or lack of response to Oncept™), should be considered. Furthermore, with somatic mutations in NRAS and PTEN being found in canine malignant melanoma(Gillard et al., 2014) similar to human melanoma hotspot sites, these may represent logical targets in the future.

The Future of Cancer Vaccines and Immunotherapy

Tumor immunology and immunotherapy is one of the most exciting and rapidly expanding fields at present. Significant resources are focused on mechanisms to simultaneously maximally stimulate an antitumor immune response while minimizing the immunosuppressive aspects of the tumor microenvironment (Catchpole et al., 2002). The recent elucidation and blockade of immunosuppressive cytokines (e.g. TGF-β, IL-10 and IL-13) and/or the negative costimulatory molecule CTLA-4(Peggs et al., 2006; Graves et al., 2009) and PD-1 (Programmed Cell Death 1 or CD279) (Callahan and Wolchok, 2013) along with the functional characterization of myeloid derived suppressor cells (MDSC) and T regulatory cells,(Biller et al., 2007; Horiuchi et al., 2009; O'Neill et al., 2009; Sherger et al., 2012) have dramatically improve cell-mediated immunity to tumors by “taking the brake” off the immune system. Immunotherapy is unlikely to become a sole modality in the treatment of cancer as the traditional modalities of surgery, radiation and/or chemotherapy are extremely likely to be used in combination with immunotherapy in the future. Like any form of anti-cancer treatment, immunotherapy appears to work best in a minimal residual disease setting, suggesting its most appropriate use will be in an adjuvant setting with local tumor therapies such as surgery and/or radiation (D.H.Thamm, 2006). Similarly, the long held belief that chemotherapy (non-corticosteroid) attenuates immune responses from cancer vaccines is beginning to be disproven through investigations on a variety of levels (Walter et al., 2006; Emens and Jaffee, 2005).

The aforementioned greatly expanded molecular understanding of the immune system has recently translated into human cancer immunotherapeutics which confer a survival benefit such as the use of the checkpoint inhibitor anti-CTLA-4 Ab, ipilimumab (Yervoy™, Bristol-Myers Squibb) and the selective BRAF inhibitors, vemurafenib (Zelboraf™, Genetech) and dabrafenib (GSK2118436, GlaxoSmithKline), in patients who are BRAF V600 mutation positive. Currently, FDA-approved Abs directed against another checkpoint inhibitor known as PD-1 receptor or against human PDL-1 (L = ligand) such as Nivolumab (Opdivo®, Bristol-Myers Squibb), Pembrolizumab (Keytruda®, Merck) and others have generated the most excitement due to ~ 20-30% of human patients having durable objective tumor responses (Wolchok et al., 2013; Callahan and Wolchok, 2013). The highest objective tumor response rates have been seen to date in patients treated with concurrent PD-1 and CTLA-4 checkpoint inhibitors (Ott et al., 2017a; Hellmann et al.,

2016; Callahan and Wolchok, 2013). Furthermore, pembrolizumab was recently given FDA approval for unresectable or metastatic, microsatellite instability-high (MSI-H) or mismatch repair deficient (dMMR) solid tumors that have progressed following prior treatment (Le et al., 2015). This is truly revolutionary as it represents the first cancer-agnostic FDA approval and is one of many recent FDA approvals based on single arm studies. There are currently over 100 different checkpoint inhibitors in development and the immuno-oncology (IO) "pendulum" has swung from the previous primarily target-dependent approaches to the current checkpoint inhibitor target-independent approach. The IO pendulum over time is highly likely to come back to the middle with a concurrent target-dependent approach (ie vaccine or similar) alongside target-independent checkpoint inhibitors (Karaki et al., 2016; Lai and Friedman, 2017)

Unfortunately, very few biomarkers of response with these clinically important agents have been found to date except for tumors which generate numerous neoantigens from tumor-specific mutations (Yuasa et al., 2017; Khagi et al., 2017). This further highlights why pembrolizumab was given FDA approval for MSI-H or dMMR solid tumors as these tumors "throw off" comparatively much higher numbers of neoantigens. A new and highly innovative personalized human cancer vaccine taking advantage of this biomarker-based approach utilizes the neoantigens specific to that individual's cancer, as they are not present in normal tissues and are highly immunogenic (Ott et al., 2017b).

Another area of extreme promise in immuno-oncology is a form of adoptive cell therapy called CAR-T (Chimeric Antigen Receptor-T cells). T cells are harvested from a patient and then genetically engineered to express a CAR on their cell surface with expansion in vitro before being reinfused back into the patient (Jackson et al., 2016). This CAR is specifically designed to recognize a specific TAA with a domain responsible for activating the T-cell when the CAR-T binds its TAA. The latest generations of CAR-T are engineered to contain important costimulatory domains that further enhance the immune response against the cancer cell containing the TAA. In 2017, an expert panel of the FDA called ODAC (Oncologic Drugs Advisory Committee) unanimously (10-0) recommended approval of CTL019 (tisagenlecleucel), an investigational CAR-T therapy utilizing CD19 as its CAR of choice for patients with B-cell acute lymphoblastic leukemia (ALL). Many CAR-T studies have found > 80-90% objective tumor responses, but side effects, including death can occur (Jackson et al., 2016). Furthermore, CAR-Ts are currently difficult to make and carry a high cost of goods to produce, making for a difficult, but not impossible commercial development path in veterinary medicine.

Checkpoints, checkpoint inhibitors and other adoptive cell transfer technologies like CAR-T and others are also starting to be better understood and pursued in veterinary diseases (Coy et al., 2017; Maekawa et al., 2016; Maekawa et al., 2017; Shosu et al., 2016; Chiku et al., 2016; Tagawa et al., 2016; Esch et al., 2013; Folkl et al., 2010; Kumar et al., 2017a; Smith et al., 2015; Anderson and Modiano, 2015; Mata et al., 2014; Mie et al., 2016; Panjwani et al., 2016; O'Connor et al., 2012; Kurupati et al., 2018; Nemoto et al., 2018; Canter et al., 2017). Furthermore, the exciting race to develop commercial veterinary specific IO therapeutics like checkpoint inhibitors and CAR-T is currently ongoing with a handful of animal health companies. It is important to review these modalities as they will potentially be used alongside cancer vaccines. As these therapeutics reduce immune tolerance and more easily generate specific anti-tumor immune responses in patients, pathologic autoimmunity was predicted and is now being seen clinically in human patients as a side effect (Ott et al., 2017a; Heinzerling and Goldinger, 2017; Kumar et al., 2017b).

In summary, the future looks extremely bright for immunotherapy. Similarly, the veterinary oncology profession is uniquely able to greatly contribute to the many advances to come in this field. Unfortunately, what works in a mouse will often not reflect the outcome in human cancer patients. Therefore, comparative immunotherapy studies utilizing veterinary patients may be able better “bridge” murine and human studies. To this end, a large number of cancers in dogs and cats appear to be remarkably stronger models for counterpart human tumors than presently available murine model systems (Khanna et al., 2009; Paoloni and Khanna, 2008; Ranieri et al., 2013; Angstadt et al., 2012; Gillard et al., 2014; LeBlanc et al., 2016a; LeBlanc et al., 2016b; Seelig et al., 2016; Fulkerson et al., 2017). This is likely due to a variety of reasons including, but not limited to, extreme similarities in the biology of the tumors (e.g. chemoresistance, radioresistance, sharing metastatic phenotypes and site selectivity, etc.), spontaneous syngeneic cancer (vs. typically an induced and/or xenogeneic cancer in murine models), and finally that the dogs and cats that are spontaneously developing these tumors are outbred, immune-competent and live in the same environment that humans do. This author ardently looks forward to the time when cancer immunotherapy plays the same significant role in the treatment and/or prevention of cancers in veterinary patients like it currently does in human cancers.

Disclosure

Dr. Bergman is a co-inventor on patent US7556805B2 and was the veterinary principal investigator for the canine melanoma vaccine Oncept® which received USDA-CVB conditional licensure in 2007 and full licensure in 2009. He also receives a minority royalty stream payment from Merial.

References

Akhtar,N., Padilla,M.L., Dickerson,E.B., Steinberg,H., Breen,M., Auerbach,R., and Helfand,S.C. (2004). Interleukin-12 inhibits tumor growth in a novel angiogenesis canine hemangiosarcoma xenograft model. Neoplasia. *6*, 106-116.

Alexander,A.N., Huelsmeyer,M.K., Mitzey,A., Dubielzig,R.R., Kurzman,I.D., Macewen,E.G., and Vail,D.M. (2006). Development of an allogeneic whole-cell tumor vaccine expressing xenogeneic gp100 and its implementation in a phase II clinical trial in canine patients with malignant melanoma. Cancer Immunol. Immunother. *55*, 433-442.

Andersen,B.M., Pluhar,G.E., Seiler,C.E., Goulart,M.R., SantaCruz,K.S., Schutten,M.M., Meints,J.P., O'Sullivan,M.G., Bentley,R.T., Packer,R.A., Thomovsky,S.A., Chen,A.V., Faissler,D., Chen,W., Hunt,M.A., Olin,M.R., and Ohlfest,J.R. (2013). Vaccination for invasive canine meningioma induces in situ production of antibodies capable of antibody-dependent cell-mediated cytotoxicity. Cancer Res. *73*, 2987-2997.

Anderson,K.L. and Modiano,J.F. (2015). Progress in Adaptive Immunotherapy for Cancer in Companion Animals: Success on the Path to a Cure. Vet Sci 2, 363-387.

Angstadt,A.Y., Thayanithy,V., Subramanian,S., Modiano,J.F., and Breen,M. (2012). A genome-wide approach to comparative oncology: high-resolution oligonucleotide aCGH of canine and human osteosarcoma pinpoints shared microaberrations. Cancer Genet. *205*, 572-587.

Autio,K.P., Ruotsalainen,J.J., Anttila,M.O., Niittykoski,M., Waris,M., Hemminki,A., Vaha-Koskela,M.J., and Hinkkanen,A.E. (2015). Attenuated Semliki Forest virus for cancer treatment in dogs: safety assessment in two laboratory Beagles. BMC. Vet Res. *11*, 170.

Beatty,P.L. and Finn,O.J. (2013). Preventing cancer by targeting abnormally expressed self-antigens: MUC1 vaccines for prevention of epithelial adenocarcinomas. Ann. N. Y. Acad. Sci *1284*, 52-56.

Bergman,P.J. (2007). Anticancer vaccines. Vet. Clin. North Am. Small Anim Pract. *37*, 1111-1119.

Bergman,P.J., Camps-Palau,M.A., McKnight,J.A., Leibman,N.F., Craft,D.M., Leung,C., Liao,J., Riviere,I., Sadelain,M., Hohenhaus,A.E., Gregor,P., Houghton,A.N., Perales,M.A., and Wolchok,J.D. (2006). Development of a xenogeneic DNA vaccine program for canine malignant melanoma at the Animal Medical Center. Vaccine *24*, 4582-4585.

Bergman,P.J., McKnight,J., Novosad,A., Charney,S., Farrelly,J., Craft,D., Wulderk,M., Jeffers,Y., Sadelain,M., Hohenhaus,A.E., Segal,N., Gregor,P., Engelhorn,M., Riviere,I., Houghton,A.N., and Wolchok,J.D. (2003). Long-term survival of dogs with advanced malignant melanoma after DNA vaccination with xenogeneic human tyrosinase: a phase I trial. Clin. Cancer Res. *9*, 1284-1290.

Biller,B.J., Elmslie,R.E., Burnett,R.C., Avery,A.C., and Dow,S.W. (2007). Use of FoxP3 expression to identify regulatory T cells in healthy dogs and dogs with cancer. Vet. Immunol. Immunopathol. *116*, 69-78.

Bird,R.C., Deinnocentes,P., Church Bird,A.E., van Ginkel,F.W., Lindquist,J., and Smith,B.F. (2011). An autologous dendritic cell canine mammary tumor hybrid-cell fusion vaccine. Cancer Immunol. Immunother. *60*, 87-97.

Bird,R.C., Deinnocentes,P., Lenz,S., Thacker,E.E., Curiel,D.T., and Smith,B.F. (2008). An allogeneic hybrid-cell fusion vaccine against canine mammary cancer. Vet. Immunol. Immunopathol. *123*, 289-304.

Boston,S.E., Lu,X., Culp,W.T., Montinaro,V., Romanelli,G., Dudley,R.M., Liptak,J.M., Mestrinho,L.A., and Buracco,P. (2014). Efficacy of systemic adjuvant therapies administered to dogs after excision of oral malignant melanomas: 151 cases (2001-2012). J Am. Vet Med. Assoc. *245*, 401-407.

Callahan,M.K. and Wolchok,J.D. (2013). At the bedside: CTLA-4- and PD-1-blocking antibodies in cancer immunotherapy. J Leukoc. Biol. *94*, 41-53.

Cangul,I.T., van Garderen,E., van der Poel,H.J., Weijer,K., and Misdorp,W. (1999). Tyrosinase gene expression in clear cell sarcoma indicates a melanocytic origin: insight from the first reported canine case. APMIS JID - 8803400 *107*, 982-988.

Canter,R.J., Grossenbacher,S.K., Foltz,J.A., Sturgill,I.R., Park,J.S., Luna,J.I., Kent,M.S., Culp,W.T.N., Chen,M., Modiano,J.F., Monjazeb,A.M., Lee,D.A., and Murphy,W.J. (2017). Radiotherapy enhances natural killer cell cytotoxicity and localization in pre-clinical canine sarcomas and first-in-dog clinical trial. J Immunother. Cancer *5*, 98.

Catchpole,B., Gould,S.M., Kellett-Gregory,L.M., and Dobson,J.M. (2002). Immunosuppressive cytokines in the regional lymph node of a dog suffering from oral malignant melanoma. J. Small Anim Pract. *43*, 464-467.

Cejalvo,T., Perise-Barrios,A.J., Del,P., I, Laborda,E., Rodriguez-Milla,M.A., Cubillo,I., Vazquez,F., Sardon,D., Ramirez,M., Alemany,R., Del,C.N., and Garcia-Castro,J. (2018). Remission of Spontaneous Canine Tumors after Systemic Cellular Viroimmunotherapy. Cancer Res. *78*, 4891-4901.

Chiku,V.M., Silva,K.L., de Almeida,B.F., Venturin,G.L., Leal,A.A., de Martini,C.C., de Rezende,E.F., Dos Santos,P.S., and de Lima,V.M. (2016). PD-1 function in apoptosis of T lymphocytes in canine visceral leishmaniasis. Immunobiology *221*, 879-888.

Chou,P.C., Chuang,T.F., Jan,T.R., Gion,H.C., Huang,Y.C., Lei,H.J., Chen,W.Y., and Chu,R.M. (2009). Effects of immunotherapy of IL-6 and IL-15 plasmids on transmissible venereal tumor in beagles. Vet. Immunol. Immunopathol.

Chu,P.Y., Pan,S.L., Liu,C.H., Lee,J., Yeh,L.S., and Liao,A.T. (2012). KIT gene exon 11 mutations in canine malignant melanoma. Vet. J.

Chuang,T.F., Lee,S.C., Liao,K.W., Hsiao,Y.W., Lo,C.H., Chiang,B.L., Lin,X.Z., Tao,M.H., and Chu,R.M. (2009). Electroporation-mediated IL-12 gene therapy in a transplantable canine cancer model. Int. J. Cancer.

Coy,J., Caldwell,A., Chow,L., Guth,A., and Dow,S. (2017). PD-1 expression by canine T cells and functional effects of PD-1 blockade. Vet Comp Oncol.

Cutrera,J., Torrero,M., Shiomitsu,K., Mauldin,N., and Li,S. (2008). Intratumoral bleomycin and IL-12 electrochemogenetherapy for treating head and neck tumors in dogs. Methods Mol. Biol. *423*, 319-325.

D.H.Thamm (2006). Interactions between radiation therapy and immunotherapy: the best of two worlds? Veterinary and Comparative Oncology *4*, 189-197.

de Vries,T.J., Smeets,M., de,G.R., Hou-Jensen,K., Brocker,E.B., Renard,N., Eggermont,A.M., van Muijen,G.N., and Ruiter,D.J. (2001). Expression of gp100, MART-1, tyrosinase, and S100 in paraffin-embedded primary melanomas and locoregional, lymph node, and visceral metastases: implications for diagnosis and immunotherapy. A study conducted by the EORTC Melanoma Cooperative Group. J Pathol. *193*, 13-20.

Denies,S., Cicchelero,L., Polis,I., and Sanders,N.N. (2016). Immunogenicity and safety of xenogeneic vascular endothelial growth factor receptor-2 DNA vaccination in mice and dogs. Oncotarget. *7*, 10905-10916.

Dickerson,E.B., Akhtar,N., Steinberg,H., Wang,Z.Y., Lindstrom,M.J., Padilla,M.L., Auerbach,R., and Helfand,S.C. (2004). Enhancement of the antiangiogenic activity of interleukin-12 by peptide targeted delivery of the cytokine to alphavbeta3 integrin. Mol. Cancer Res. *2*, 663-673.

Dickerson,E.B., Fosmire,S., Padilla,M.L., Modiano,J.F., and Helfand,S.C. (2002). Potential to target dysregulated interleukin-2 receptor expression in canine lymphoid and hematopoietic malignancies as a model for human cancer. J. Immunother. *25*, 36-45.

Dow,S., Elmslie,R., Kurzman,I., MacEwen,G., Pericle,F., and Liggitt,D. (2005). Phase I study of liposome-DNA complexes encoding the interleukin-2 gene in dogs with osteosarcoma lung metastases. Hum. Gene Ther. *16*, 937-946.

Emens,L.A. and Jaffee,E.M. (2005). Leveraging the activity of tumor vaccines with cytotoxic chemotherapy. Cancer Res. *65*, 8059-8064.

Epple,L.M., Bemis,L.T., Cavanaugh,R.P., Skope,A., Mayer-Sonnenfeld,T., Frank,C., Olver,C.S., Lencioni,A.M., Dusto,N.L., Tal,A., Har-Noy,M., Lillehei,K.O., Katsanis,E., and Graner,M.W. (2013). Prolonged remission of advanced bronchoalveolar adenocarcinoma in a dog treated with autologous, tumour-derived chaperone-rich cell lysate (CRCL) vaccine. Int. J Hyperthermia *29*, 390-398.

Esch,K.J., Juelsgaard,R., Martinez,P.A., Jones,D.E., and Petersen,C.A. (2013). Programmed death 1-mediated T cell exhaustion during visceral leishmaniasis impairs phagocyte function. J Immunol. *191*, 5542-5550.

Fazekas,J., Furdos,I., Singer,J., and Jensen-Jarolim,E. (2016). Why man's best friend, the dog, could also benefit from an anti-HER-2 vaccine. Oncol. Lett. *12*, 2271-2276.

Finocchiaro,L.M., Fiszman,G.L., Karara,A.L., and Glikin,G.C. (2008). Suicide gene and cytokines combined nonviral gene therapy for spontaneous canine melanoma. Cancer Gene Ther. *15*, 165-172.

Finocchiaro,L.M., Fondello,C., Gil-Cardeza,M.L., Rossi,U.A., Villaverde,M.S., Riveros,M.D., and Glikin,G.C. (2015). Cytokine-Enhanced Vaccine and Interferon-beta plus Suicide Gene Therapy as Surgery Adjuvant Treatments for Spontaneous Canine Melanoma. Hum. Gene Ther. *26*, 367-376.

Finocchiaro,L.M. and Glikin,G.C. (2008). Cytokine-enhanced vaccine and suicide gene therapy as surgery adjuvant treatments for spontaneous canine melanoma. Gene Ther. *15*, 267-276.

Finocchiaro,L.M. and Glikin,G.C. (2012). Cytokine-enhanced vaccine and suicide gene therapy as surgery adjuvant treatments for spontaneous canine melanoma: 9 years of follow-up. Cancer Gene Ther. *19*, 852-861.

Finocchiaro,L.M.E., Spector,A.I.M., Agnetti,L., Arbe,M.F., and Glikin,G.C. (2018). Combination of Suicide and Cytokine Gene Therapies as Surgery Adjuvant for Canine Mammary Carcinoma. Vet Sci *5*.

Folkl,A., Wen,X., Kuczynski,E., Clark,M.E., and Bienzle,D. (2010). Feline programmed death and its ligand: characterization and changes with feline immunodeficiency virus infection. Vet Immunol. Immunopathol. *134*, 107-114.

Fulkerson,C.M., Dhawan,D., Ratliff,T.L., Hahn,N.M., and Knapp,D.W. (2017). Naturally Occurring Canine Invasive Urinary Bladder Cancer: A Complementary Animal Model to Improve the Success Rate in Human Clinical Trials of New Cancer Drugs. Int. J Genomics *2017*, 6589529.

Gabai,V., Venanzi,F.M., Bagashova,E., Rud,O., Mariotti,F., Vullo,C., Catone,G., Sherman,M.Y., Concetti,A., Chursov,A., Latanova,A., Shcherbinina,V., Shifrin,V., and Shneider,A. (2014). Pilot study of p62 DNA vaccine in dogs with mammary tumors. Oncotarget. *5*, 12803-12810.

Gavazza,A., Lubas,G., Fridman,A., Peruzzi,D., Impellizeri,J.A., Luberto,L., Marra,E., Roscilli,G., Ciliberto,G., and Aurisicchio,L. (2013). Safety and efficacy of a genetic vaccine targeting telomerase plus chemotherapy for the therapy of canine B-cell lymphoma. Hum. Gene Ther. *24*, 728-738.

Gibson,H.M., Veenstra,J.J., Jones,R., Vaishampayan,U., Sauerbrey,M., Bepler,G., Lum,L., Reyes,J., Weise,A., and Wei,W.Z. (2015). Induction of HER2 Immunity in Outbred Domestic Cats by DNA Electrovaccination. Cancer Immunol. Res. *3*, 777-786.

Gillard,M., Cadieu,E., De,B.C., Abadie,J., Vergier,B., Devauchelle,P., Degorce,F., Dreano,S., Primot,A., Dorso,L., Lagadic,M., Galibert,F., Hedan,B., Galibert,M.D., and Andre,C. (2014). Naturally occurring melanomas in dogs as models for non-UV pathways of human melanomas. Pigment Cell Melanoma Res. *27*, 90-102.

Goubier,A., Fuhrmann,L., Forest,L., Cachet,N., Evrad-Blanchard,M., Juillard,V., and Fischer,L. (2008). Superiority of needle-free transdermal plasmid delivery for the induction of antigen-specific IFNgamma T cell responses in the dog. Vaccine *26*, 2186-2190.

Gradilone,A., Gazzaniga,P., Ribuffo,D., Bottoni,U., Frati,L., Aglian,A.M., Sorvillo,V., Piperno,A., Scuderi,N., and Cigna,E. (2012). Prognostic significance of tyrosinase expression in sentinel lymph node biopsy for ultra-thin, thin, and thick melanomas. Eur. Rev. Med. Pharmacol. Sci *16*, 1367-1376.

Graves,S.S., Stone,D., Loretz,C., Peterson,L., McCune,J.S., Mielcarek,M., and Storb,R. (2009). Establishment of long-term tolerance to SRBC in dogs by recombinant canine CTLA4-Ig. Transplantation *88*, 317-322.

Grosenbaugh,D.A., Leard,A.T., Bergman,P.J., Klein,M.K., Meleo,K., Susaneck,S., Hess,P.R., Jankowski,M.K., Jones,P.D., Leibman,N.F., Johnson,M.H., Kurzman,I.D., and Wolchok,J.D. (2011). Safety and efficacy of a xenogeneic DNA vaccine encoding for human tyrosinase as adjunctive treatment for oral malignant melanoma in dogs following surgical excision of the primary tumor. Am. J. Vet. Res. *72*, 1631-1638.

Gyorffy,S., Rodriguez-Lecompte,J.C., Woods,J.P., Foley,R., Kruth,S., Liaw,P.C., and Gauldie,J. (2005). Bone marrow-derived dendritic cell vaccination of dogs with naturally occurring melanoma by using human gp100 antigen. J. Vet. Intern. Med. *19*, 56-63.

Hampel,V., Schwarz,B., Kempf,C., Kostlin,R., Schillinger,U., Kuchenhoff,H., Fenske,N., Brill,T., and Hirschberger,J. (2007). Adjuvant immunotherapy of feline fibrosarcoma with recombinant feline interferon-omega. J. Vet. Intern. Med. *21*, 1340-1346.

Heinzerling,L. and Goldinger,S.M. (2017). A review of serious adverse effects under treatment with checkpoint inhibitors. Curr. Opin. Oncol. *29*, 136-144.

Hellmann,M.D., Friedman,C.F., and Wolchok,J.D. (2016). Combinatorial Cancer Immunotherapies. Adv. Immunol. *130*, 251-277.

Hernandez-Granados,A.J., Franco-Molina,M.A., Coronado-Cerda,E.E., Zapata-Benavides,P., Gamboa,E.M., Ramos-Zayas,Y., Santana-Krimskaya,S.E., and Rodriguez-Padilla,C. (2018). Immunogenic potential of three transmissible venereal tumor cell lysates to prime canine-dendritic cells for cancer immunotherapy. Res. Vet Sci *121*, 23-30.

Herzog,A., Buchholz,J., Ruess-Melzer,K., Lang,J., and Kaser-Hotz,B. (2013). [Concurrent irradiation and DNA tumor vaccination in canine oral malignant melanoma: a pilot study]. Schweiz. Arch. Tierheilkd. *155*, 135-142.

Hogge,G.S., Burkholder,J.K., Culp,J., Albertini,M.R., Dubielzig,R.R., Yang,N.S., and Macewen,E.G. (1999). Preclinical development of human granulocyte-macrophage colony-stimulating factor-transfected melanoma cell vaccine using established canine cell lines and normal dogs. Cancer Gene Ther. *6*, 26-36.

Hoopes,P.J., Moodie,K.L., Petryk,A.A., Petryk,J.D., Sechrist,S., Gladstone,D.J., Steinmetz,N.F., Veliz,F.A., Bursey,A.A., Wagner,R.J., Rajan,A., Dugat,D., Crary-Burney,M., and Fiering,S.N. (2017). Hypo-fractionated Radiation, Magnetic Nanoparticle Hyperthermia and a Viral Immunotherapy Treatment of Spontaneous Canine Cancer. Proc. SPIE Int. Soc. Opt. Eng *10066*.

Hoopes,P.J., Wagner,R.J., Duval,K., Kang,K., Gladstone,D.J., Moodie,K.L., Crary-Burney,M., Ariaspulido,H., Veliz,F.A., Steinmetz,N.F., and Fiering,S.N. (2018). Treatment of Canine Oral Melanoma with Nanotechnology-Based Immunotherapy and Radiation. Mol. Pharm. *15*, 3717-3722.

Horiuchi,Y., Tominaga,M., Ichikawa,M., Yamashita,M., Jikumaru,Y., Nariai,Y., Nakajima,Y., Kuwabara,M., and Yukawa,M. (2009). Increase of regulatory T cells in the peripheral blood of dogs with metastatic tumors. Microbiol. Immunol. *53*, 468-474.

Impellizeri,J.A., Ciliberto,G., and Aurisicchio,L. (2012). Electro-gene-transfer as a new tool for cancer immunotherapy in animals. Vet Comp Oncol.

Impellizeri,J.A., Gavazza,A., Greissworth,E., Crispo,A., Montella,M., Ciliberto,G., Lubas,G., and Aurisicchio,L. (2018). Tel-eVax: a genetic vaccine targeting telomerase for treatment of canine lymphoma. J Transl. Med. *16*, 349.

Jackson,H.J., Rafiq,S., and Brentjens,R.J. (2016). Driving CAR T-cells forward. Nat. Rev. Clin. Oncol. *13*, 370-383.

Jahnke,A., Hirschberger,J., Fischer,C., Brill,T., Kostlin,R., Plank,C., Kuchenhoff,H., Krieger,S., Kamenica,K., and Schillinger,U. (2007). Intra-tumoral gene delivery of feIL-2, feIFN-gamma and feGM-CSF using magnetofection as a neoadjuvant treatment option for feline fibrosarcomas: a phase-I study. J. Vet. Med. A Physiol Pathol. Clin. Med. *54*, 599-606.

Johnston,K.B., Monteiro,J.M., Schultz,L.D., Chen,L., Wang,F., Ausensi,V.A., Dell,E.C., Santos,E.B., Moore,R.A., Palker,T.J., Stanley,M.A., and Jansen,K.U. (2005). Protection of beagle dogs from mucosal challenge with canine oral papillomavirus by immunization with recombinant adenoviruses expressing codon-optimized early genes. Virology *336*, 208-218.

Jourdier,T.M., Moste,C., Bonnet,M.C., Delisle,F., Tafani,J.P., Devauchelle,P., Tartaglia,J., and Moingeon,P. (2003). Local immunotherapy of spontaneous feline fibrosarcomas using recombinant poxviruses expressing interleukin 2 (IL2). Gene Ther. *10*, 2126-2132.

Junco,J.A., Basalto,R., Fuentes,F., Bover,E., Reyes,O., Pimentel,E., Calzada,L., Castro,M.D., Arteaga,N., Lopez,Y., Hernandez,H., Bringas,R., Garay,H., Peschke,P., Bertot,J., and Guillen,G. (2008). Gonadotrophin releasing hormone-based vaccine, an effective candidate for prostate cancer and other hormone-sensitive neoplasms. Adv. Exp. Med. Biol. *617*, 581-587.

Kamstock,D., Elmslie,R., Thamm,D., and Dow,S. (2007). Evaluation of a xenogeneic VEGF vaccine in dogs with soft tissue sarcoma. Cancer Immunol. Immunother. *56*, 1299-1309.

Kamstock,D., Guth,A., Elmslie,R., Kurzman,I., Liggitt,D., Coro,L., Fairman,J., and Dow,S. (2006). Liposome-DNA complexes infused intravenously inhibit tumor angiogenesis and elicit antitumor activity in dogs with soft tissue sarcoma. Cancer Gene Ther. *13*, 306-317.

Karaki,S., Anson,M., Tran,T., Giusti,D., Blanc,C., Oudard,S., and Tartour,E. (2016). Is There Still Room for Cancer Vaccines at the Era of Checkpoint Inhibitors. Vaccines. (Basel) *4*.

Khagi,Y., Kurzrock,R., and Patel,S.P. (2017). Next generation predictive biomarkers for immune checkpoint inhibition. Cancer Metastasis Rev. *36*, 179-190.

Khanna,C., Anderson,P.M., Hasz,D.E., Katsanis,E., Neville,M., and Klausner,J.S. (1997). Interleukin-2 liposome inhalation therapy is safe and effective for dogs with spontaneous pulmonary metastases. Cancer *79*, 1409-1421.

Khanna,C., Hasz,D.E., Klausner,J.S., and Anderson,P.M. (1996). Aerosol delivery of interleukin 2 liposomes is nontoxic and biologically effective: canine studies. Clin. Cancer Res. *2*, 721-734.

Khanna,C., London,C., Vail,D., Mazcko,C., and Hirschfeld,S. (2009). Guiding the optimal translation of new cancer treatments from canine to human cancer patients. Clin. Cancer Res. *15*, 5671-5677.

Kruth,S.A. (1998). Biological response modifiers: interferons, interleukins, recombinant products, liposomal products. Vet. Clin. North Am. Small Anim Pract. *28*, 269-295.

Kumar,S.R., Kim,D.Y., Henry,C.J., Bryan,J.N., Robinson,K.L., and Eaton,A.M. (2017a). Programmed death ligand 1 is expressed in canine B cell lymphoma and downregulated by MEK inhibitors. Vet Comp Oncol.

Kumar,V., Chaudhary,N., Garg,M., Floudas,C.S., Soni,P., and Chandra,A.B. (2017b). Current Diagnosis and Management of Immune Related Adverse Events (irAEs) Induced by Immune Checkpoint Inhibitor Therapy. Front Pharmacol. *8*, 49.

Kuntsi-Vaattovaara,H., Verstraete,F.J.M., Newsome,J.T., and Yuan,H. (2003). Resolution of persistent oral papillomatosis in a dog after treatment with a recombinant canine oral papillomavirus vaccine. Veterinary and Comparative Oncology *1*, 57-63.

Kurupati,R.K., Zhou,X., Xiang,Z., Keller,L.H., and Ertl,H.C.J. (2018). Safety and immunogenicity of a potential checkpoint blockade vaccine for canine melanoma. Cancer Immunol. Immunother. *67*, 1533-1544.

Kyte,J.A., Mu,L., Aamdal,S., Kvalheim,G., Dueland,S., Hauser,M., Gullestad,H.P., Ryder,T., Lislerud,K., Hammerstad,H., and Gaudernack,G. (2006). Phase I/II trial of melanoma therapy with dendritic cells transfected with autologous tumor-mRNA. Cancer Gene Ther. *13*, 905-918.

Lai,X. and Friedman,A. (2017). Combination therapy of cancer with cancer vaccine and immune checkpoint inhibitors: A mathematical model. PLoS. One. *12*, e0178479.

Le,D., Uram,J., Wang,H., Bartlett,B., Kemberling,H., Eyring,A., Skora,A., Luber,B., Azad,N., Laheru,D., Biedrzycki,B., Donehower,R., Zaheer,A., Fisher,G., Crocenzi,T., Lee,J., Duffy,S., Goldberg,R., de la Chapelle,A., Koshiji,M., Bhaijee,F., Huebner,T., Hruban,R., Wood,L., Cuka,N., Pardoll,D., Papadopoulos,N., Kinzler,K., Zhou,S., Cornish,T., Taube,J., Anders,R., Eshleman,J., Vogelstein,B., and Diaz,L. (2015). PD-1 Blockade in Tumors with Mismatch-Repair Deficiency. N. Engl. J Med. *372*, 2509-2520.

LeBlanc,A.K., Breen,M., Choyke,P., Dewhirst,M., Fan,T.M., Gustafson,D.L., Helman,L.J., Kastan,M.B., Knapp,D.W., Levin,W.J., London,C., Mason,N., Mazcko,C., Olson,P.N., Page,R., Teicher,B.A., Thamm,D.H., Trent,J.M., Vail,D.M., and Khanna,C. (2016a). Perspectives from man's best friend: National Academy of Medicine's Workshop on Comparative Oncology. Sci Transl. Med. *8*, 324ps5.

LeBlanc,A.K., Mazcko,C., Brown,D.E., Koehler,J.W., Miller,A.D., Miller,C.R., Bentley,R.T., Packer,R.A., Breen,M., Boudreau,C.E., Levine,J.M., Simpson,R.M., Halsey,C., Kisseberth,W., Rossmeisl,J.H., Jr., Dickinson,P.J., Fan,T.M., Corps,K.,

Aldape,K., Puduvalli,V., Pluhar,G.E., and Gilbert,M.R. (2016b). Creation of an NCI comparative brain tumor consortium: informing the translation of new knowledge from canine to human brain tumor patients. Neuro. Oncol. *18*, 1209-1218.

Liao,J.C., Gregor,P., Wolchok,J.D., Orlandi,F., Craft,D., Leung,C., Houghton,A.N., and Bergman,P.J. (2006). Vaccination with human tyrosinase DNA induces antibody responses in dogs with advanced melanoma. Cancer Immun. *6*, 8.

Maekawa,N., Konnai,S., Okagawa,T., Nishimori,A., Ikebuchi,R., Izumi,Y., Takagi,S., Kagawa,Y., Nakajima,C., Suzuki,Y., Kato,Y., Murata,S., and Ohashi,K. (2016). Immunohistochemical Analysis of PD-L1 Expression in Canine Malignant Cancers and PD-1 Expression on Lymphocytes in Canine Oral Melanoma. PLoS. One. *11*, e0157176.

Maekawa,N., Konnai,S., Takagi,S., Kagawa,Y., Okagawa,T., Nishimori,A., Ikebuchi,R., Izumi,Y., Deguchi,T., Nakajima,C., Kato,Y., Yamamoto,K., Uemura,H., Suzuki,Y., Murata,S., and Ohashi,K. (2017). A canine chimeric monoclonal antibody targeting PD-L1 and its clinical efficacy in canine oral malignant melanoma or undifferentiated sarcoma. Sci Rep. *7*, 8951.

Manley,C.A., Leibman,N.F., Wolchok,J.D., Riviere,I.C., Bartido,S., Craft,D.M., and Bergman,P.J. (2011). Xenogeneic Murine Tyrosinase DNA Vaccine for Malignant Melanoma of the Digit of Dogs. J. Vet. Intern. Med. *25*, 94-99.

Marconato,L., Frayssinet,P., Rouquet,N., Comazzi,S., Leone,V.F., Laganga,P., Rossi,F., Vignoli,M., Pezzoli,L., and Aresu,L. (2013). Randomized, placebo-controlled, double-blinded chemo-immunotherapy clinical trial in a Pet Dog model of Diffuse Large B-cell Lymphoma. Clin. Cancer Res.

Marconato,L., Frayssinet,P., Rouquet,N., Comazzi,S., Leone,V.F., Laganga,P., Rossi,F., Vignoli,M., Pezzoli,L., and Aresu,L. (2014). Randomized, placebo-controlled, double-blinded chemoimmunotherapy clinical trial in a pet dog model of diffuse large B-cell lymphoma. Clin. Cancer Res. *20*, 668-677.

Marconato,L., Stefanello,D., Sabattini,S., Comazzi,S., Riondato,F., Laganga,P., Frayssinet,P., Pizzoni,S., Rouquet,N., and Aresu,L. (2015). Enhanced therapeutic effect of APAVAC immunotherapy in combination with dose-intense chemotherapy in dogs with advanced indolent B-cell lymphoma. Vaccine *33*, 5080-5086.

Mason,N.J., Coughlin,C.M., Overley,B., Cohen,J.N., Mitchell,E.L., Colligon,T.A., Clifford,C.A., Zurbriggen,A., Sorenmo,K.U., and Vonderheide,R.H. (2008). RNA-loaded CD40-activated B cells stimulate antigen-specific T-cell responses in dogs with spontaneous lymphoma. Gene Ther. *15*, 955-965.

Mason,N.J., Gnanandarajah,J.S., Engiles,J.B., Gray,F., Laughlin,D., Gaurnier-Hausser,A., Wallecha,A., Huebner,M., and Paterson,Y. (2016). Immunotherapy with a HER2-Targeting Listeria Induces HER2-Specific Immunity and Demonstrates Potential Therapeutic Effects in a Phase I Trial in Canine Osteosarcoma. Clin. Cancer Res. *22*, 4380-4390.

Mata,M., Vera,J., Gerken,C., Rooney,C.M., Miller,T., Pfent,C., Wang,L.L., Wilson-Robles,H.M., and Gottschalk,S. (2014). Towards immunotherapy with redirected T cells in a large animal model: Ex vivo activation, expansion, and genetic modification of canine T cells. J Immunother. *37*, 407-415.

McLean,J.L. and Lobetti,R.G. (2015). Use of the melanoma vaccine in 38 dogs: The South African experience. J S. Afr. Vet Assoc. *86*, 1246.

Mie,K., Shimada,T., Akiyoshi,H., Hayashi,A., and Ohashi,F. (2016). Change in peripheral blood lymphocyte count in dogs following adoptive immunotherapy using lymphokine-activated T killer cells combined with palliative tumor resection. Vet Immunol. Immunopathol. *177*, 58-63.

Milner,R.J., Salute,M., Crawford,C., Abbot,J.R., and Farese,J. (2006). The immune response to disialoganglioside GD3 vaccination in normal dogs: a melanoma surface antigen vaccine. Vet. Immunol. Immunopathol. *114*, 273-284.

Morse,M.A., Mosca,P.J., Clay,T.M., and Lyerly,H.K. (2002). Dendritic cell maturation in active immunotherapy strategies. Expert. Opin. Biol. Ther. *2*, 35-43.

Murakami,A., Mori,T., Sakai,H., Murakami,M., Yanai,T., Hoshino,Y., and Maruo,K. (2011). Analysis of KIT expression and KIT exon 11 mutations in canine oral malignant melanomas. Vet. Comp Oncol. *9*, 219-224.

Musser,M.L., Berger,E.P., Parsons,C., Kathariou,S., and Johannes,C.M. (2019). Vaccine strain Listeria monocytogenes abscess in a dog: a case report. BMC. Vet. Res. *15*, 467.

Nemoto,Y., Shosu,K., Okuda,M., Noguchi,S., and Mizuno,T. (2018). Development and characterization of monoclonal antibodies against canine PD-1 and PD-L1. Vet Immunol. Immunopathol. *198*, 19-25.

O'Connor,C.M., Sheppard,S., Hartline,C.A., Huls,H., Johnson,M., Palla,S.L., Maiti,S., Ma,W., Davis,R.E., Craig,S., Lee,D.A., Champlin,R., Wilson,H., and Cooper,L.J. (2012). Adoptive T-cell therapy improves treatment of canine non-Hodgkin lymphoma post chemotherapy. Sci Rep. 2, 249.

O'Neill,K., Guth,A., Biller,B., Elmslie,R., and Dow,S. (2009). Changes in regulatory T cells in dogs with cancer and associations with tumor type. J. Vet. Intern. Med. *23*, 875-881.

Okano,F. and Yamada,K. (2000). Canine interleukin-18 induces apoptosis and enhances Fas ligand mRNA expression in a canine carcinoma cell line. Anticancer Res. *20*, 3411-3415.

Ott,P.A., Hodi,F.S., Kaufman,H.L., Wigginton,J.M., and Wolchok,J.D. (2017a). Combination immunotherapy: a road map. J Immunother. Cancer *5*, 16.

Ott,P.A., Hu,Z., Keskin,D.B., Shukla,S.A., Sun,J., Bozym,D.J., Zhang,W., Luoma,A., Giobbie-Hurder,A., Peter,L., Chen,C., Olive,O., Carter,T.A., Li,S., Lieb,D.J., Eisenhaure,T., Gjini,E., Stevens,J., Lane,W.J., Javeri,I., Nellaiappan,K., Salazar,A.M., Daley,H., Seaman,M., Buchbinder,E.I., Yoon,C.H., Harden,M., Lennon,N., Gabriel,S., Rodig,S.J., Barouch,D.H., Aster,J.C., Getz,G., Wucherpfennig,K., Neuberg,D., Ritz,J., Lander,E.S., Fritsch,E.F., Hacohen,N., and Wu,C.J. (2017b). An immunogenic personal neoantigen vaccine for patients with melanoma. Nature *547*, 217-221.

Ottnod,J.M., Smedley,R.C., Walshaw,R., Hauptman,J.G., Kiupel,M., and Obradovich,J.E. (2013). A retrospective analysis of the efficacy of Oncept vaccine for the adjunct treatment of canine oral malignant melanoma. Vet Comp Oncol. *11*, 219-229.

Panjwani,M.K., Smith,J.B., Schutsky,K., Gnanandarajah,J., O'Connor,C.M., Powell,D.J., Jr., and Mason,N.J. (2016). Feasibility and Safety of RNA-transfected CD20-specific Chimeric Antigen Receptor T Cells in Dogs with Spontaneous B Cell Lymphoma. Mol. Ther. *24*, 1602-1614.

Paoloni,M. and Khanna,C. (2008). Translation of new cancer treatments from pet dogs to humans. Nat. Rev. Cancer *8*, 147-156.

Peggs,K.S., Quezada,S.A., Korman,A.J., and Allison,J.P. (2006). Principles and use of anti-CTLA4 antibody in human cancer immunotherapy. Curr. Opin. Immunol. *18*, 206-213.

Perales,M.A., Yuan,J., Powel,S., Gallardo,H.F., Rasalan,T.S., Gonzalez,C., Manukian,G., Wang,J., Zhang,Y., Chapman,P.B., Krown,S.E., Livingston,P.O., Ejadi,S., Panageas,K.S., Engelhorn,M.E., Terzulli,S.L., Houghton,A.N., and Wolchok,J.D. (2008). Phase I/II study of GM-CSF DNA as an adjuvant for a multipeptide cancer vaccine in patients with advanced melanoma. Mol. Ther. *16*, 2022-2029.

Peruzzi,D., Mesiti,G., Ciliberto,G., La,M.N., and Aurisicchio,L. (2010). Telomerase and HER-2/neu as targets of genetic cancer vaccines in dogs. Vaccine *28*, 1201-1208.

Phillips, J. C., Blackford, J. T., Lembcke, L. M., Grosenbaugh, D. A., and Leard, A. T. Evaluation of Needle-free Injection Devices for Intramuscular Vaccination in Horses. J Eq Vet Sci 31, 738-743. 2011.

Ref Type: Generic

Phillips,J.C., Lembcke,L.M., Noltenius,C.E., Newman,S.J., Blackford,J.T., Grosenbaugh,D.A., and Leard,A.T. (2012). Evaluation of tyrosinase expression in canine and equine melanocytic tumors. Am. J. Vet. Res. *73*, 272-278.

Piras,L.A., Riccardo,F., Iussich,S., Maniscalco,L., Gattino,F., Martano,M., Morello,E., Lorda,M.S., Rolih,V., Garavaglia,F., De,M.R., Lardone,E., Collivignarelli,F., Mignacca,D., Giacobino,D., Ferrone,S., Cavallo,F., and Buracco,P. (2017). Prolongation of survival of dogs with oral malignant melanoma treated by en bloc surgical resection and adjuvant CSPG4-antigen electrovaccination. Vet Comp Oncol. *15*, 996-1013.

Quintin-Colonna,F., Devauchelle,P., Fradelizi,D., Mourot,B., Faure,T., Kourilsky,P., Roth,C., and Mehtali,M. (1996). Gene therapy of spontaneous canine melanoma and feline fibrosarcoma by intratumoral administration of histoincompatible cells expressing human interleukin-2. Gene Ther. *3*, 1104-1112.

Ramos-Vara,J.A., Beissenherz,M.E., Miller,M.A., Johnson,G.C., Pace,L.W., Fard,A., and Kottler,S.J. (2000). Retrospective study of 338 canine oral melanomas with clinical, histologic, and immunohistochemical review of 129 cases. Vet Pathol JID - 0312020 *37*, 597-608.

Ramos-Vara,J.A. and Miller,M.A. (2011). Immunohistochemical identification of canine melanocytic neoplasms with antibodies to melanocytic antigen PNL2 and tyrosinase: comparison with Melan A. Vet. Pathol. *48*, 443-450.

Ranieri,G., Gadaleta,C.D., Patruno,R., Zizzo,N., Daidone,M.G., Hansson,M.G., Paradiso,A., and Ribatti,D. (2013). A model of study for human cancer: Spontaneous occurring tumors in dogs. Biological features and translation for new anticancer therapies. Crit Rev. Oncol. Hematol. *88*, 187-197.

Regan,D., Guth,A., Coy,J., and Dow,S. (2016). Cancer immunotherapy in veterinary medicine: Current options and new developments. Vet J *207*, 20-28.

Riccardo,F., Iussich,S., Maniscalco,L., Lorda,M.S., La,R.G., Arigoni,M., De,M.R., Gattino,F., Lanzardo,S., Lardone,E., Martano,M., Morello,E., Prestigio,S., Fiore,A.,

Quaglino,E., Zabarino,S., Ferrone,S., Buracco,P., and Cavallo,F. (2014). CSPG4-specific immunity and survival prolongation in dogs with oral malignant melanoma immunized with human CSPG4 DNA. Clin. Cancer Res. *20*, 3753-3762.

Rodriguez-Lecompte,J.C., Kruth,S., Gyorffy,S., Wan,Y.H., and Gauldie,J. (2004). Cell-based cancer gene therapy: breaking tolerance or inducing autoimmunity? Anim Health Res. Rev. *5*, 227-234.

Sanchez,D., Cesarman-Maus,G., Amador-Molina,A., and Lizano,M. (2018). Oncolytic Viruses for Canine Cancer Treatment. Cancers. (Basel) *10*.

Sarbu,L., Kitchell,B.E., and Bergman,P.J. (2015). Safety of administering the canine melanoma DNA vaccine (Oncept) to cats with malignant melanoma - a retrospective study. J Feline. Med. Surg.

Seelig,D.M., Avery,A.C., Ehrhart,E.J., and Linden,M.A. (2016). The Comparative Diagnostic Features of Canine and Human Lymphoma. Vet Sci *3*.

Seymour,L., Bogaerts,J., Perrone,A., Ford,R., Schwartz,L.H., Mandrekar,S., Lin,N.U., Litiere,S., Dancey,J., Chen,A., Hodi,F.S., Therasse,P., Hoekstra,O.S., Shankar,L.K., Wolchok,J.D., Ballinger,M., Caramella,C., and de Vries,E.G. (2017). iRECIST: guidelines for response criteria for use in trials testing immunotherapeutics. Lancet Oncol. *18*, e143-e152.

Shahabi,V., Seavey,M.M., Maciag,P.C., Rivera,S., and Wallecha,A. (2011). Development of a live and highly attenuated Listeria monocytogenes-based vaccine for the treatment of Her2/neu-overexpressing cancers in human. Cancer Gene Ther. *18*, 53-62.

Sherger,M., Kisseberth,W., London,C., Olivo-Marston,S., and Papenfuss,T.L. (2012). Identification of myeloid derived suppressor cells in the peripheral blood of tumor bearing dogs. BMC. Vet Res. *8*, 209.

Shosu,K., Sakurai,M., Inoue,K., Nakagawa,T., Sakai,H., Morimoto,M., Okuda,M., Noguchi,S., and Mizuno,T. (2016). Programmed Cell Death Ligand 1 Expression in Canine Cancer. In Vivo *30*, 195-204.

Siddiqui,F., Li,C.Y., Zhang,X., LaRue,S.M., Dewhirst,M.W., Ullrich,R.L., and Avery,P.R. (2006). Characterization of a recombinant adenovirus vector encoding heat-inducible feline interleukin-12 for use in hyperthermia-induced gene-therapy. Int. J. Hyperthermia *22*, 117-134.

Skubitz,K.M. and Anderson,P.M. (2000). Inhalational interleukin-2 liposomes for pulmonary metastases: a phase I clinical trial. Anticancer Drugs *11*, 555-563.

Smedley,R.C., Lamoureux,J., Sledge,D.G., and Kiupel,M. (2011). Immunohistochemical diagnosis of canine oral amelanotic melanocytic neoplasms. Vet. Pathol. *48*, 32-40.

Smith,J.B., Panjwani,M.K., Schutsky,K., Gnanandarajah,J., Calhoun,S., Cooper,L., Mason,N., and Powell,D.J. (2015). Feasibility and safety of cCD20 RNA CAR-bearing T cell therapy for the treatment of canine B cell malignancies. J Immunother. Cancer *3*, 123.

Sorenmo,K.U., Krick,E., Coughlin,C.M., Overley,B., Gregor,T.P., Vonderheide,R.H., and Mason,N.J. (2011). CD40-activated B cell cancer vaccine improves second clinical remission and survival in privately owned dogs with non-Hodgkin's lymphoma. PLoS. One. *6*, e24167.

Suckow,M.A. (2013). Cancer vaccines: harnessing the potential of anti-tumor immunity. Vet J *198*, 28-33.

Tagawa,M., Maekawa,N., Konnai,S., and Takagi,S. (2016). Evaluation of Costimulatory Molecules in Peripheral Blood Lymphocytes of Canine Patients with Histiocytic Sarcoma. PLoS. One. *11*, e0150030.

Tamura,K., Arai,H., Ueno,E., Saito,C., Yagihara,H., Isotani,M., Ono,K., Washizu,T., and Bonkobara,M. (2007). Comparison of dendritic cell-mediated immune responses among canine malignant cells. J. Vet. Med. Sci. *69*, 925-930.

Tamura,K., Yamada,M., Isotani,M., Arai,H., Yagihara,H., Ono,K., Washizu,T., and Bonkobara,M. (2008). Induction of dendritic cell-mediated immune responses against canine malignant melanoma cells. Vet. J. *175*, 126-129.

Tateyama,S., Priosoeryanto,B.P., Yamaguchi,R., Uchida,K., Ogiwara,K., and Suchiya,A.T. (1995). In vitro growth inhibition activities of recombinant feline interferon on all lines derived from canine tumors. Res. Vet. Sci. *59*, 275-277.

Thacker,E.E., Nakayama,M., Smith,B.F., Bird,R.C., Muminova,Z., Strong,T., Timares,L., Korokhov,N., O'Neill,A.M., de Gruijl,T.D., Glasgow,J.N., Tani,K., and Curiel,D.T. (2009). A genetically engineered adenovirus vector targeted to CD40 mediates transduction of canine dendritic cells and promotes antigen-specific immune responses in vivo. Vaccine.

Treggiari,E., Grant,J.P., and North,S.M. (2016). A retrospective review of outcome and survival following surgery and adjuvant xenogeneic DNA vaccination in 32 dogs with oral malignant melanoma. J Vet Med. Sci *78*, 845-850.

Turek,M.M., Thamm,D.H., Mitzey,A., Kurzman,I.D., Huelsmeyer,M.K., Dubielzig,R.R., and Vail,D.M. (2007). Human granulocyte and macrophage colony-stimulating factor DNA cationic-lipid complexed autologous tumour cell vaccination in the treatment of canine B-cell multicentric lymphoma. Veterinary and Comparative Oncology *5*, 219-231.

U'Ren,L.W., Biller,B.J., Elmslie,R.E., Thamm,D.H., and Dow,S.W. (2007). Evaluation of a novel tumor vaccine in dogs with hemangiosarcoma. J. Vet. Intern. Med. *21*, 113-120.

Verganti,S., Berlato,D., Blackwood,L., Amores-Fuster,I., Polton,G.A., Elders,R., Doyle,R., Taylor,A., and Murphy,S. (2017). Use of Oncept melanoma vaccine in 69 canine oral malignant melanomas in the UK. J Small Anim Pract. *58*, 10-16.

von,E.H., Sadeghi,A., Carlsson,B., Rivera,P., Loskog,A., Segall,T., Korsgren,O., and Totterman,T.H. (2008). Efficient adenovector CD40 ligand immunotherapy of canine malignant melanoma. J. Immunother. *31*, 377-384.

Walter,C.U., Biller,B.J., Lana,S.E., Bachand,A.M., and Dow,S.W. (2006). Effects of chemotherapy on immune responses in dogs with cancer. J. Vet. Intern. Med. *20*, 342-347.

Weir,C., Hudson,A.L., Moon,E., Ross,A., Alexander,M., Peters,L., Langova,V., Clarke,S.J., Pavlakis,N., Davey,R., and Howell,V.M. (2014). Streptavidin: a novel immunostimulant for the selection and delivery of autologous and syngeneic tumor vaccines. Cancer Immunol. Res. *2*, 469-479.

Weir,C., Oksa,A., Millar,J., Alexander,M., Kynoch,N., Walton-Weitz,Z., Mackenzie-Wood,P., Tam,F., Richards,H., Naylor,R., Cheng,K., Bennett,P., Petrovsky,N., and Allavena,R. (2018). The Safety of an Adjuvanted Autologous Cancer Vaccine Platform in Canine Cancer Patients. Vet Sci *5*.

Whitley,E.M., Bird,A.C., Zucker,K.E., and Wolfe,L.G. (1995). Modulation by canine interferon-gamma of major histocompatibility complex and tumor-associated antigen expression in canine mammary tumor and melanoma cell lines. Anticancer Res. *15*, 923-929.

Wolchok,J.D., Kluger,H., Callahan,M.K., Postow,M.A., Rizvi,N.A., Lesokhin,A.M., Segal,N.H., Ariyan,C.E., Gordon,R.A., Reed,K., Burke,M.M., Caldwell,A., Kronenberg,S.A., Agunwamba,B.U., Zhang,X., Lowy,I., Inzunza,H.D., Feely,W., Horak,C.E., Hong,Q., Korman,A.J., Wigginton,J.M., Gupta,A., and Sznol,M. (2013). Nivolumab plus ipilimumab in advanced melanoma. N. Engl. J Med. *369*, 122-133.

Wolchok,J.D., Yuan,J., Houghton,A.N., Gallardo,H.F., Rasalan,T.S., Wang,J., Zhang,Y., Ranganathan,R., Chapman,P.B., Krown,S.E., Livingston,P.O., Heywood,M., Riviere,I., Panageas,K.S., Terzulli,S.L., and Perales,M.A. (2007).

Safety and immunogenicity of tyrosinase DNA vaccines in patients with melanoma. Mol. Ther. *15*, 2044-2050.

Yamaguchi,T. and Sakaguchi,S. (2006). Regulatory T cells in immune surveillance and treatment of cancer. Semin. Cancer Biol. *16*, 115-123.

Yannelli,J.R., Wouda,R., Masterson,T.J., Avdiushko,M.G., and Cohen,D.A. (2016). Development of an autologous canine cancer vaccine system for resectable malignant tumors in dogs. Vet Immunol. Immunopathol. *182*, 95-100.

Yu,W.Y., Chuang,T.F., Guichard,C., El-Garch,H., Tierny,D., Laio,A.T., Lin,C.S., Chiou,K.H., Tsai,C.L., Liu,C.H., Li,W.C., Fischer,L., and Chu,R.M. (2011). Chicken HSP70 DNA vaccine inhibits tumor growth in a canine cancer model. Vaccine *29*, 3489-3500.

Yuan,J., Ku,G.Y., Gallardo,H.F., Orlandi,F., Manukian,G., Rasalan,T.S., Xu,Y., Li,H., Vyas,S., Mu,Z., Chapman,P.B., Krown,S.E., Panageas,K., Terzulli,S.L., Old,L.J., Houghton,A.N., and Wolchok,J.D. (2009). Safety and immunogenicity of a human and mouse gp100 DNA vaccine in a phase I trial of patients with melanoma. Cancer Immun. *9*, 5.

Yuasa,T., Masuda,H., Yamamoto,S., Numao,N., and Yonese,J. (2017). Biomarkers to predict prognosis and response to checkpoint inhibitors. Int. J Clin. Oncol. *22*, 629-634.

Zagury,D. and Gallo,R.C. (2004). Anti-cytokine Ab immune therapy: present status and perspectives. Drug Discov. Today *9*, 72-81.

Zuleger,C.L., Kang,C., Ranheim,E.A., Kurzman,I.D., Macklin,M.D., Newton,M.A., Wolchok,J.D., Vail,D.M., Eriksson,E., and Albertini,M.R. (2017). Pilot study of safety and feasibility of DNA microseeding for treatment of spontaneous canine melanoma. Vet Med. Sci *3*, 134-145.

Chapter 3

Bovine Mucosal Vaccines: Challenges and Opportunities

Paola Elizalde[1] and Philip J. Griebel[1,2*]

[1]VIDO-Intervac, University of Saskatchewan, Saskatoon, SK Canada
[2]School of Public Health, University of Saskatchewan, Saskatoon, SK Canada

*Corresponding Author: philip.griebel@usask.cca

DOI: https://doi.org/10.21775/9781913652593.03

Abstract

Mucosal vaccines have been used in cattle for almost 50 years with intranasal (Todd et al, 1971) and oral (Mebus et al. 1972) delivery being primary routes of vaccine delivery (reviewed in Hodgins et al. 2005). There is increasing interest in the use of mucosal vaccination in cattle for several reasons. In newborn calves, intranasal vaccines provide a strategy to reduce vaccine interference by maternal antibody and enhance disease protection as maternal antibody wanes. Mucosal vaccines also offer an advantage in controlling both clinical disease and reducing transmission of mucosal pathogens. As our understanding of mucosal immunity improves, mucosal vaccines may offer opportunities to activate both innate and adaptive mucosal effector cells and improve control of mucosal infections while maintaining mucosal barrier integrity and vital mucosal functions. Furthermore, a greater understanding of host-microbiome interactions may inform vaccine strategies to control opportunistic pathogens residing within the commensal microbiome. Unique aspects of the bovine mucosal immune system are considered within the context of vaccination and the limitations and potential benefits of mucosal vaccines and vaccine delivery systems are discussed. Finally, the potential for new vaccine delivery vehicles and vaccination strategies to improve mucosa vaccine efficacy are discussed within the context of opportunities to address current infectious disease challenges in cattle.

Introduction

Vaccines provide an effective tool for the control of numerous infectious diseases in ruminants and parenteral injection remains the primary route of vaccine delivery in cattle. There are, however, an increasing number of commercial vaccines delivered mucosally in cattle (Table 1) and interest in mucosal vaccination is increasing. There are a number of important infectious agents and disease complexes that have not been effectively controlled with parenteral vaccines. For example, maternal antibody interference limits the efficacy of many modified-live viral (MLV) vaccines injected parenterally in young calves (Chamorro et al. 2016). The efficacy of bacterial vaccines for the control of bovine respiratory disease has also been equivocal following parenteral vaccination of calves experiencing the stress of weaning, transportation, and social mixing (Larson and Step, 2012). Persistent mucosal infections, such as *Mycobacterium avium* subpsecies *paratuberculosis* (MAP), have also proven an elusive target for development of an effective vaccine (Barkema et al. 2017). Thus, there has been considerable interest in mucosal vaccine delivery, especially when the pathogens of interest infect mucosal surfaces (Lawan et al. 2018) or mucosal delivery provides a strategy to circumvent vaccine interference by maternal antibody (Hill et al. 2012). This review of mucosal vaccines will focus on both past successes and new opportunities and challenges for mucosal vaccination in cattle. New opportunities for the use of mucosal vaccination are being created by a greater understanding of the development and function of the bovine mucosal

Table 1. Commercial Vaccines Used for Mucosal Immunization of Cattle[1]

Pathogen	Vaccine Type	Vaccine Delivery Route	
		Oral	Nasal
Bovine coronavirus	ML[2]	√	√
Bovine herpesvirus-1	ML		√
Bovine respiratory syncytial virus	ML		√
Bovine rotavirus	ML	√	
Dictyocaulus viviparus	Infectious larvae	√	
Parainfluenza-3	ML		√
Mannheimia haemolytica	ML		√
Pasteurella multocida	ML		√

[1]Based on information available for North America

[2]ML: modified-live pathogen

immune system (Osman et al. 2018) and a better understanding of host-microbial interactions at mucosal surfaces (Taschuk and Griebel, 2012). Further, new vaccine delivery technologies are emerging to address the major challenges associated with mucosal vaccination. These challenges are diverse and include low efficiency of antigen delivery to mucosal surfaces, rapid degradation of mucosal vaccine antigens, limited immunogenicity of vaccine delivery systems, and the need to induce both humoral and cellular immunity (Holmgren and Czerkinsky, 2005; Pavot et al. 2012). The potential benefits of mucosal vaccination are substantial, however, in light of the fact that over 90% of pathogens invade through mucosal surfaces (McGhee and Mestecky, 1990).

Comparison of Immune Responses Induced by Parenteral Versus Mucosal Vaccination

Immune Responses Induced by Parenteral Vaccination

Most commercial vaccines given to cattle are administered by either subcutaneous (SC) or intramuscular (IM) injection and these vaccines contain not only an antigenic fraction but may be formulated with preservatives, antimicrobials and immune stimulating compounds, referred to as adjuvants. Adjuvants perform a variety of functions but a major function is to induce the release of either endogenous danger signals or to function directly as an exogenous danger signal recognized by pattern recognition receptors (PRRs) (Burakova et al. 2018). Activation of PRRs initiates a local inflammatory response at the vaccine injection site that recruits a variety of phagocytic cells. Dendritic cells (DC) are present throughout all tissues, are a functionally and phenotypically diverse population of cells in cattle (Fries and Griebel, 2011) and are abundant within skin and the underlying subcutis (Mirabzadeh-Ardakani et al. 2016). DCs play a key role in the uptake and transport of vaccine antigens from the injection site by migrating through lymph vessels to the nearest draining lymph node (Figure 1). Within the draining lymph node, DCs present vaccine antigens to naïve T_{helper} (Th) and cytotoxic T lymphocytes (CTL) to initiate effector and memory immune responses that contribute to vaccine-induced cell-mediated immune protection (Figure 1).

Humoral or antibody responses are also an important component of immune protection induced by parenteral vaccination. Germinal Centers (GC) located in draining lymph nodes function as the site for clonal expansion and selection of B cells expressing high-affinity antibody receptors specific for vaccine antigens (Figure 1). Parenteral vaccination results in the induction of primarily IgG secreting plasma cells and IgG isotype-switched memory B cells but vaccine adjuvants can alter the magnitude of antibody production and bias the IgG antibody isotypes produced following vaccination (Ioannou et al. 2002). While parenteral vaccination induces primarily IgG antibody that circulates in blood, there may be enhanced production of IgA antibody at mucosal surfaces when using specific vaccine adjuvants (Gaertner et al. 2009) or modified-live viral vaccines (Kolb et al. 2020). Enhanced IgA responses in the upper respiratory tract (URT) were observed following parenteral injection of a modified-live bovine respiratory syncytial virus (BRSV), a virus with a tropism for both the upper and lower respiratory tract (Brodersen, 2010). Further research is required to determine whether induction of an IgA response in the URT was the result of limited viral spread from the SC injection site to the respiratory tract (Walz et al. 2017) or possibly the homing of effector cells from the lymph node draining the vaccine injection site.

The expansion of specific effector and memory T and B cells is the ultimate goal of vaccination but numerous studies have characterized the important role regulatory T cells (Tregs) play in control of acquired immune responses (reviewed in Belkaid and Tarbell, 2009). There is increasing interest in the role Tregs may play in constraining vaccine-induced immune responses and developing strategies to circumvent this Treg activity (Berod et al. 2012). Tregs have been identified in cattle (Roussey et al. 2014) but there are no studies specifically analyzing Treg activity following parenteral vaccination of cattle. Therefore, it remains to be determined what role Tregs may play in modulating either the magnitude or duration of bovine immune responses following parenteral vaccination (Figure 1).

Immune responses induced by mucosal vaccination

Mucosal vaccines interact with the immune system in ways distinct from the interactions that occur between parenteral vaccines and the systemic immune system. First, a mucosal vaccine, such as MLV vaccines, may interact directly with mucosal epithelium, a key component of the mucosal immune system. Mucosal epithelial cells play a direct role in recognizing foreign antigens and initiating inflammation and cell recruitment (Peterson and Artis, 2014). Epithelial cell inflammatory responses result in the recruitment and activation of DC

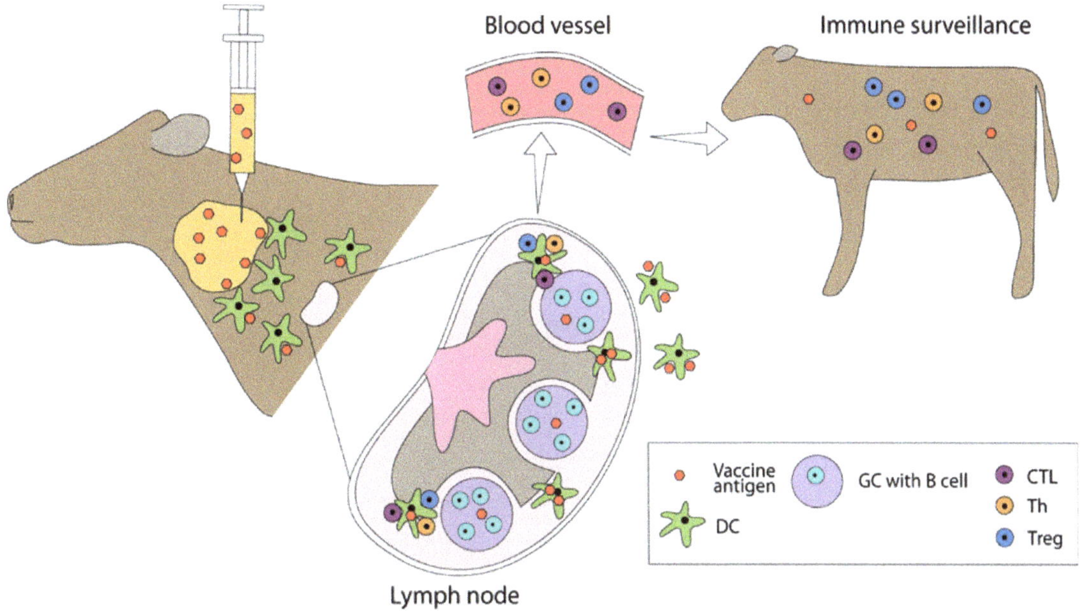

Figure 1. Induction of an immune response following a parenteral vaccine injection. Vaccines induce a local inflammatory response, which recruits dendritic cells (DC), macrophages and neutrophils to the injection site. DCs play a key role in the uptake of vaccine antigens and delivering these antigens to the draining lymph node. Antigen processing and presentation by DCs is essential for the activation and clonal expansion of naïve $CD4^+$ T helper (Th) cells and CD8+ cytotoxic T lymphocytes (CTL). Effective vaccine responses result in the release of both memory and effector T cells from the lymph node that can then circulate in blood and provide immune surveillance throughout the body. Subsequent encounters with vaccine antigen or the target pathogens can reactivate memory cells. Soluble vaccine antigen can also interact with B cells in germinal centres (GC) to induce B cell activation and differentiation into B cell memory and antibody secreting plasma cells. The Th cells induced by parenteral vaccination regulate plasma cell development, resulting in production of primarily IgG antibody. Further investigations are required to determine whether parenteral vaccination induces regulatory T cells (Tregs) that may constrain the magnitude and duration of immune responses.

subpopulations that can acquire vaccine antigens and migrate through lymphatics to the draining lymph nodes where the induction of immune responses occurs (Figure 2). Secondly, mucosal vaccines and vaccine delivery vehicles may be taken-up by mucosa-associated lymphoid tissues (MALT) through microfold (M)-cell trancytosis.

Mucosal surfaces of the bovine eye, nasopharynx, oral cavity, gastro-intestinal tract, and reproductive tract have specialized organized MALT that function as immune induction sites (reviewed in Liebler-Tenorio and Pabst, 2006). Bovine MALT develops during gestation but maturation of both structure and function occurs during the first week of life following colonization of mucosal surfaces by commensal micro-organisms (Liang et al. 2014; Osman et al. 2018). Thus, MALT has the capacity to sample and respond to both commensal micro-organisms and vaccine antigens within the first week of life (Osman et al. 2017).

Within MALT and mucosa draining lymph nodes, DCs present vaccine antigens to naïve Th and CTL to initiate effector and memory immune responses that contribute to vaccine-induced cell-mediated immune protection (Figure 2). Follicular DCs also acquire and retain antigens that induce the development of vaccine specific B cells within lymphoid follicles located in the tonsils (Figure 2) and other MALT. Effector and memory T cells and IgA secreting plasma cells induced within MALT express the specific homing receptor, $\alpha_4\beta_7$, and chemokine receptors (Kunkel and Butcher, 2002) which guide trafficking of these effector cells back to the mucosal surfaces where immune induction occurred (Figure 2). This selective homing of effector cells is critical for T cell immune surveillance of mucosal surfaces and ensures IgA secreting plasma cells localize at mucosal surfaces. A close association between IgA plasma cells and mucosal epithelium facilitates IgA transport by the polymeric immunoglobulin receptors (pIgR) expressed within epithelial cells (Johansen and Kaetzel, 2011). Rapid transport of secretory (S)IgA across epithelial surfaces results in elevated levels of IgA at mucosal surfaces, making it essential to collect mucosal secretions when monitoring IgA antibody responses following mucosal vaccination (Figure 2).

The mucosal immune system plays a dual function in maintaining the integrity of mucosal barriers throughout the body. First, the adaptive immune system generates antigen-specific effector cells and IgA antibody that prevent attachment and invasion of both commensal micro-organisms and pathogens that would disrupt the integrity of the mucosal barrier. Second, the mucosal immune system generates effector cells that regulate inflammatory responses induced by micro-organisms, toxins, or dietary macromolecules that could compromise tissue integrity and function (reviewed in McCarville and Ayres 2018). Both innate and adaptive immune cells are involved in

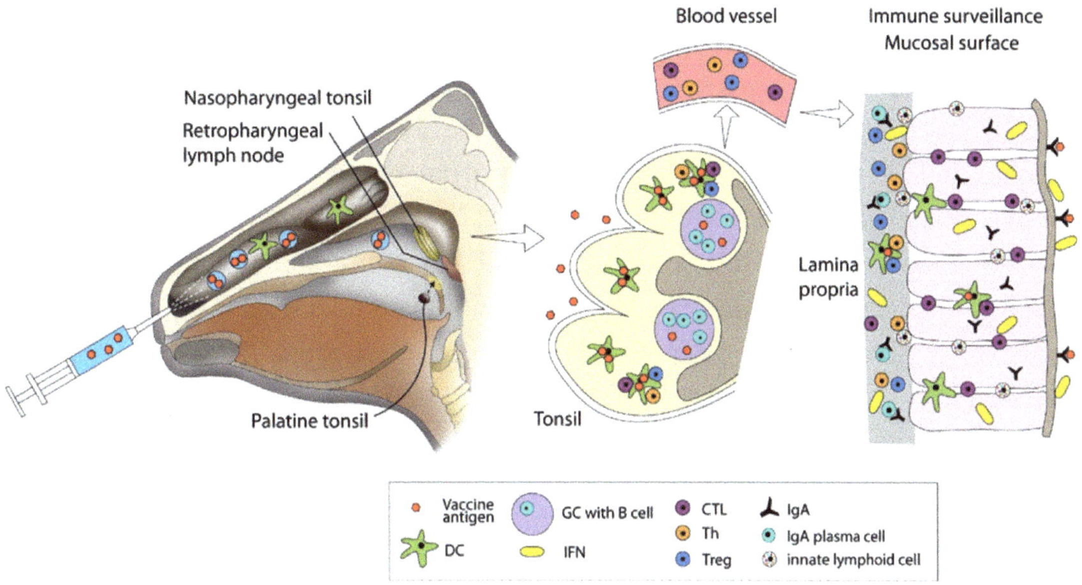

Figure 2. Induction of a mucosal immune response following intranasal vaccination. Modified-live viral vaccines may interact directly with mucosal epithelium or be trancytosed by M-cells overlying organized mucosal-associated lymphoid tissues, such as the nasopharyngeal and palatine tonsils. Viral replication induces a local inflammatory response, which recruits dendritic cells (DC), macrophages and neutrophils to the site of viral infection and replication. DCs take up and process vaccine antigens and present these antigens either in the tonsils or deliver antigens to lymph nodes draining the upper respiratory tract, such as the retropharyngeal lymph node. The initial encounter between the vaccine virus and mucosal epithelium, innate lymphoid cells, and DCs may result in local interferon (IFN) production which enhances immune surveillance and DC activation. Antigen presentation by DCs is essential for activation and clonal expansion of naïve $CD4^{+}$ T helper (Th) cells and $CD8^{+}$ cytotoxic T lymphocytes (CTL). Effective vaccine responses result in the release of both memory and effector T cells from the tonsils and lymph nodes that circulate in blood before homing back to mucosal tissue, providing local immune surveillance. Subsequent encounters with vaccine antigen or the target pathogens can reactivate memory cells. Soluble vaccine antigen also interact with B cells in germinal centres (GC) to induce B cell activation and differentiation into B cell memory and antibody secreting plasma cells. The Th cells induced by intranasal vaccination regulate plasma cell development, resulting in production of primarily IgA antibody. IgA secreting plasma cells localize beneath the mucosal epithelium and dimeric IgA istransported across the mucosal epithelium where it can bind pathogens, preventing infection of epithelial cells and the shedding of infectious viral particles. Regulatory T cells (Tregs) are abundant at mucosal surfaces but further investigation is required to understand their role in constraining immune responses induced by mucosal vaccination.

regulating inflammatory responses at mucosal surfaces. In cattle, both regulatory T (Roussey et al. 2016) and B cells (Jimbo et al. Manuscript submitted) have been identified in mucosal tissues and MALT. Recruitment of non-conventional CD335$^+$ T cells to the URT mucosa following a primary bovine herpesvrues-1 infection also suggests innate lymphoid cells (ILCs) play an important role in bovine mucosal immune responses (Osman and Griebel, 2017). Bovine non-conventional CD335$^+$ T cells produce interferon (IFN)γ (Connelley et al. 2014) and further studies are required to determine if these or other types of ILCs contribute to the IFN response observed following IN vaccination with some MLV vaccines (Figure 2). Further investigations are also required to understand the role regulatory T and B cells and ILCs may play in both the innate and adaptive immune responses induced by mucosal vaccination (Figure 2).

In conclusion, parenteral and mucosal vaccines differ greatly not only in route of delivery but also in interactions that occur with components of the bovine immune system (Table I). Activation of the systemic or peripheral immune system by parenteral vaccines and activation of the mucosal immune system following mucosal vaccine delivery results in the generation of distinct effector cell and antibody responses. This functional dichotomy is most apparent in the humoral immune system (Table 1) but may also be as distinct within cell-mediated immune responses as our understanding increases regarding the effect of vaccination on both innate and adaptive regulatory cells. Furthermore, imprinting of effector and memory cell homing in the systemic and mucosal immune induction sites results in separate immune surveillance capacities following parenteral versus mucosal immunization. Understanding effector cell trafficking has important implications when selecting vaccines to prevent infections by pathogens that localize to mucosal surfaces versus pathogens that invade mucosal surfaces but subsequently disseminate throughout the body. Mucosal vaccination may provide the best defenses against pathogen invasion but may not control disease caused by a pathogen that evades mucosal defences and infects internal organs. Alternatively, parenteral vaccines that prevent or minimize clinical disease caused by infection of internal tissues or organs may not limit transmission of infectious agents shed in secretions from mucosal surfaces. Therefore, the decision to use mucosal versus parenteral vaccination may need to consider both protection of an individual animal from disease and limiting spread of a pathogen within a population.

Table 2. Comparison of mucosal and systemic immune systems and immune responses induced following parenteral and mucosal vaccination.

Mucosal Vaccines and Mucosal Immunity	Parenteral Vaccines and Systemic Immunity
Vaccine Delivery Route Ocular Intranasal Oral Vaginal	**Vaccine Delivery Route** Intradermal Subcutaneous Intramuscular Intravenous
Lymphoid Tissues Tonsils Peyer's patches Solitary lymphoid follicles Lamina propira leukocytes Lymph nodes draining mucosa	**Lymphoid Tissue** Blood Spleen Lymph nodes draining skin, muscle, CNS, and internal organs
Effector Mechanisms Antibody - secretory IgA, T cells – Cytotoxic, helper, regulatory Innate Lymphoid Cells Cytokines	**Effector Mechanisms** Antibody – IgM, IgG, IgE T cells – Cytotoxic, helper, regulatory Innate Lymphoid Cells Cytokines

Mucosal Vaccines Targeting Bovine Viral Respiratory Pathogens

Induction of Adaptive Immune Responses to Viral Pathogens

Despite the use of vaccines, bovine respiratory disease (BRD) persists as a major cause of disease and economic losses in the global cattle industry and is prevalent in both young dairy calves and 5 to 6 month old beef calves when weaned, transported, and commingled in large groups (Hoerlein and Marsh,1957; Miles, 2009; Step et al. 2008). Several factors limit the efficacy of vaccines targeting BRD pathogens. First, BRD is a disease complex caused by multiple viral pathogens with bovine herpesvirus-1 (BHV-1), bovine respiratory syncytial virus (BRSV), bovine viral diarrhea virus (BVDV)-1 and BVDV-2, and parainfluenza (PI)-3 virus all implicated as causal agents. Emerging viral pathogens, such as bovine coronavirus and bovine influenza may also contribute to BRD. Primary viral respiratory infections alter pulmonary defences and allow URT comnmensal bacteria, such as *Mannheima haemolytica* (Yates et al. 1983) and *Mycoplasma bovis* (Prysliak et al. 2011) to colonize the lung and cause secondary bacterial infections. The failure to eliminate BRD with vaccination may also be due, in part, to vaccination practices. Many beef calves receive their first vaccination for BRD pathogens on arrival in the feedlot, which coincides with the period of highest risk for respiratory infections (Ribble et al.

1995; Duff and Galyean, 2007). Commingling weaned calves from multiple sources increases exposure to multiple BRD pathogens and there is limited time for vaccines to induce protective immunity when vaccination occurs after weaning. Most vaccines currently licensed for BRD are injected parenterally, either SC or IM, whereas viruses causing BRD are often inhaled and infect both the upper and lower respiratory tract. Systemic immune responses induced by parenteral MLV vaccines can reduce BRD clinical disease but have limited capacity to prevent viral infections, such as BHV-1, in the upper respiratory tract (URT) when primary vaccination is 5 days prior to viral challenge (Woolums et al. 2003; Fairbanks et al. 2004).

Early calfhood vaccination is one strategy now used to establish immune responses prior to the risk of respiratory viral infection but maternal antibody can interfere with the immunogenicity of parenteral MLV vaccines (Chamorro et al. 2016). For example, Ellis et al. (2001) reported parenteral vaccination with a modified-live BVDV-2 vaccine had no significant impact on decay of maternal antibody and there was no evidence of immune memory following a booster vaccination. These data provided evidence that neutralizing maternal antibody prevented the parenteral MLV vaccine from generating sufficient viral protein to induce detectable antibody responses. Intranasal (IN) vaccination of young calves with viral pathogens has provided strong evidence, however, that both antibody and cell-mediated immune responses can be induced when maternal antibody is present. Endsley et al. (2004) reported IN delivery of BVDV-2 to young calves with maternal antibody developed detectable virus-specific T cell responses and immune memory that was protective after maternal antibody was no longer present (Ridpath et al. 2003). BVDV may not provide a typical example of immune responses induced by a MLV vaccine delivered IN since BVDV causes a viremia and spreads systemically.

More recent studies with multivalent MLV vaccines have confirmed, however, that IN vaccination of neonatal calves can induce immune responses that are protective against multiple respiratory viruses. Xue et al. (2010) vaccinated neonatal calves IN with a multivalent, MLV vaccine containing 5 respiratory viruses and subsequent viral challenges confirmed protection against clinical disease. Disease protection was evaluated within three to 4 weeks after immunization, indicating rapid onset of protective immunity but no studies were performed to evaluate duration of immunity. This study demonstrated the capacity of the mucosal immune system in the URT to respond to multiple vaccine antigens when calves were vaccinated at less than 8 days of age. Calves used in this study were reared without colostrum so the question remained whether maternal antibody could interfere with mucosal vaccination.

Subsequent studies analyzed immune responses and disease protection following IN vaccination of seropositive neonatal calves with multivalent, MLV vaccines. These studies confirmed maternal IgG was not secreted in nasal secretions following transfer of high levels of maternal IgG specific for viruses used to immunize newborn calves IN (Hill et al. 2012). Further, while maternal IgA specific for respiratory vaccine viruses was present in nasal secretions this IgA was no longer detectable within 5 to 7 days after birth. These results are consistent with previous studies demonstrating less than 10% of maternal antibody is of the IgA isotype (Korhonen et al. 2000) and that little IgG is present in bovine nasal secretions (March and Pradhud, 1971). In the same IN vaccine study, it was confirmed that a MLV vaccine induced local IgA antibody responses to multiple vaccine viruses even when maternal antibody was present (Hill et al. 2012). A significant increase in IgA production was detected in nasal secretions within 10 to 14 days after vaccination but IgA levels began to decline within 5 weeks after vaccination. A booster vaccination after a 5-week interval induced anamnestic responses with a further increase in IgA titres specific for individual vaccine viruses (Hill et al. 2012). This observation is important since it revealed that local IgA antibody induced by IN vaccination did not prevent induction of an anamnestic IgA response when boosting with the same vaccine.

The study by Hill et al. (2012) did not confirm the vaccine-induced IgA antibody responses were protective and raised questions regarding short duration of immunity and immune memory following IN vaccination of newborn calves. This question is important since maternal antibody was reported to interfere with the induction of protective immunity when neonatal calves were immunized IN with a MLV vaccine, with BRSV as one component, and calves were subsequently challenged with BRSV (Ellis et al. 2010). The significance of these findings may be limited since the MLV vaccine used for IN immunization was a commercial vaccine formulated for SC delivery. Further, calves were challenged with BRSV 4.5 months after the single IN immunization. Thus, no studies have directly compared the magnitude or duration of immune responses induced, in the absence or presence of maternal antibody, when using vaccines formulated for mucosal delivery. The possibility remains that maternal antibody may impact immune responses induced by mucosal vaccination and this issue should be addressed for each vaccine used in young calves. Maternal antibody interference could occur with MLV vaccines or vectored vaccines that replicate at sites such as the lung where IgG antibody may be present.

Duration of immune protection and immune memory following IN vaccination of neonatal calves has been addressed in several studies. Ellis et el. (2013) reported a single IN vaccination of seropositive neonatal calves with a MLV vaccine induced significant immune protection against BRSV challenge at 7 but not 14 weeks post-vaccination. These observations suggest a booster vaccination is required to maintain protective immunity over a period of three or more months. Consistent with this observation, Hill et al. (2019) reported a single IN vaccination of seropositive neonatal calves with a multivalent MLV vaccine resulted in BHV-1 specific immune memory that persisted for at least 5 months. This immune memory was not sufficient to significantly reduce fever, weight loss, or virus shedding following aerosol challenge with virulent BHV-1 but was sufficient to reduce fatal secondary bacterial pneumonia. Immune memory was sufficient at 5 months post-vaccination to support rapid onset of protective immunity within 4 days after an IN booster vaccination. Immune protection following the booster vaccination was significantly better than that observed following a primary IN vaccination given to age-matched cohorts. Therefore, primary IN vaccination of seropositive neonatal calves, combined with a second IN vaccination several months later, provides a strategy to minimize the risk of BRD when calves are weaned and transported to a feedlot (Hill et al. 2019).

The study by Hill et al. (2019) also addressed the issue of potential interactions between the systemic and mucosal immune system when delivering vaccines mucosally versus parenterally. Neonatal calves that received a primary IN vaccination with a MLV vaccine then received a booster vaccination 5 months later with the same vaccine injected either IN or SC. Immune responses, clinical disease, and virus shedding were monitored following BHV-1 challenge. The SC and IN booster vaccinations were equally effective in reducing fever, weight loss and fatal secondary bacterial pneumonias. A rapid and sustained reduction in BHV-1 shedding in nasal secretions was, however, only observed with IN and not SC booster vaccination. Further, increased levels of viral-specific IgA antibody were observed in nasal secretions of calves receiving the IN but not the SC booster vaccination. Thus,

SC vaccinated calves appeared clinically healthy with no fever or weight loss but shed high levels of BHV-1 in their nasal secretions (Hill et al. 2019). These observations are consistent with known differences in the trafficking of effector and memory T and B cells when lymphocytes are activated in MALT versus peripheral lymph nodes (reviewed in Wiedle et al. 2001). The capacity of IN vaccination to significantly increase IgA secretion and reduce virus shedding in nasal secretions highlights a major advantage of mucosal versus parenteral vaccination. When disease transmission involves a mucosal surface then mucosal vaccination is better able to protect both the individual animal and reduce the risk of disease transmission within a population.

Mucosal vaccination for BRD viral pathogens is effective in mature cattle (Smith et al. 1978) but use is limited for a variety of reasons. Some BRD pathogens, such as BHV-1 and BVDV-1 and -2, are also important causes of fetal infection following viremia in pregnant cows. IN immunization with a BHV-1 vaccine can prevent fetal infection (Smith et al. 1978) but systemic immunity, with high levels of neutralizing IgG antibody in blood, protects well against fetal infection (Cortese et al. 1998; Xue et al. 2011). Furthermore, parenteral vaccination enhances the transfer of maternal IgG antibody specific for BRD viral pathogens to newborn calves (Hill et al. 2012). Therefore, few studies have analyzed the immunogenicity of IN vaccines for BRD in older animals. A recent study compared IgA antibody responses induced by IN vaccination of dairy cows with a multivalent MLV vaccine, containing BHV-1 as one viral component, and cows were vaccinated either 2 weeks pre-partum or the day after calving (Cortese et al. 2017). The increase in BHV-1 specific nasal IgA titres during the two weeks post-vaccination was significantly greater when cows were vaccinated post-partum. This study indicates pregnancy can have a negative impact on mucosal immune responses induced by IN vaccination. Transcriptome analysis of blood leukocyte responses following parenteral vaccination of cows during both the pre- and post-partum period suggested suppression of immune function also occurs during the post-partum period (Weikardet al. 2015). Thus, further studies are required to determine whether IgA responses to mucosal vaccination are suppressed during the early post-natal period. A comparison of IgA responses in cows vaccinated IN immediately post-partum and later during the lactation cycle might reveal whether both mucosal and systemic immunity are equally affected during this transition period in pregnant cows.

The effect of stress on immune responses induced by mucosal vaccination is also highly relevant to vaccines used to control BRD in young calves since weaning and transportation can double fatal pneumonia following experimental BHV-1 and *Mannheimia haemolytica* infections (Hodgson et al. 2012). MLV IN vaccines are used to immunize weaned beef calves following their arrival in the feedlot and this practice is based both on rapid onset of protective immunity within 4 days of vaccination as well as activation of innate immune defences in the URT (Hill et al. 2019). Studies have not directly compared the efficacy or immunogenicity of IN vaccines in calves when stress is either present or absent. The capacity of IN vaccines to induce protective immunity in stressed calves has been evaluated, however, by immunizing seronegative beef calves the day after weaning and transportation (Hill et al. 2019). These studies revealed primary IN vaccination did not significantly reduce clinical disease but significantly reduced viral shedding when calves were challenged with BHV-1 four days after vaccination (Hill et al. 2019). A significantly better reduction in clinical disease and virus shedding was observed when calves receiving a booster versus primary IN vaccination. Thus, mucosal immune memory was able to support a protective anamnestic immune response despite the presence of stressors known to elevate serum cortisol levels (Hodgson et al. 2012). As indicated previously, further studies are required in both old and young animals to clearly determine how stress impacts mucosal immune responses. Different stressors or combinations of stressors have significantly different effects on immune responses observed following infection by a respiratory virus, such as BHV-1 (Griebel et al. 2014). Therefore, it would not be surprising if mucosal vaccine immunogenicity and efficacy were reduced when animals experience concurrent stressors.

Induction of Innate Immune Responses by Modified-Live Viral Vaccines

IN vaccination with MLV vaccines containing attenuated bovine respiratory viruses induces both acquired and innate immune responses (Todd et al. 1971; Savan et al. 1979). Characterization of mucosal vaccine induced innate immune responses has focused primarily on the analysis of interferon (IFN) production. Infection of the bovine URT with BHV-1 (Todd et al. 1972) and PI-3 (Rosenquist and Loan, 1967) increased IFN production in nasal secretions. In these early studies, IFN responses were detected using biological assays to measure inhibition of viral replication in tissue culture. Thus, the specific types of IFN present in nasal secretions were not identified. Three distinct bovine IFN gene families have been identified, including Type I (IFNα and IFNβ), II (IFNγ), and III (IFNλ), and all three IFN types are known to have antiviral activity. Recent studies confirm IFNα, IFNβ and IFNγ protein secretion and gene transcription increase in the URT following BHV-1 infection (Osman et al. 2017). Associated with this IFN response was a significant increase in the expression of a variety of interferon-stimulated genes, including genes with known anti-viral effector functions. Thus, it is not surprising IN vaccination with MLV vaccines, containing attenuated BHV-1 and PI-3, can reduce infection by unrelated respiratory viruses (Todd et al. 1972). Furthermore, studies with recombinant bovine IFNα have shown IFN also has a significant protective effect when BHV-1 infected cattle are challenged with a bacterial respiratory pathogen (Czarniecki et al. 1985). Activation of innate immune defences following IN delivery of a MLV vaccine is one reason IN delivery is preferred over parenteral injection when vaccinating calves in the face of a BRD outbreak that could involve multiple viral pathogens.

IFNα and IFNγ levels in nasal secretions are usually below the level of detection in healthy calves (Hodgson et al. 2012, Osman et al. 2017). Following IN vaccination of BHV-1 seropositive neonatal calves with MLV vaccines, containing BHV-1 and PI3, there is a significant increase in both IFNα and IFNγ protein levels in nasal secretions for 5 to 7 days post-vaccination (Midla et al. In Press). The levels of IFNα and IFNγ induced by IN vaccination were, however, approximately 40-fold lower than that observed following experimental BHV-1 infection of seronegative 6 month old calves (Osman et al. 2017). This low level of vaccine-induced IFN secretion may reflect a lack of immune competence in neonatal animals or the biological properties of attenuated vaccine viral strains. Primary IN vaccination of 6 month-old calves with a MLV vaccine induced much higher levels of both IFNα and IFNγ in nasal secretions (Hill et al. 2019). The study by Midla et al. (In Press) directly compared two different commercial MLV vaccines and found significant differences between vaccines when comparing IFNα concentrations in nasal secretions and the fraction of vaccinated animals responding with detectable increases in IFN secretion. These data indicate MLV vaccines may differ significantly in their capacity to activate innate immune responses.

Activation of innate immune responses by IN vaccines has also been analyzed in older animals. It is interesting to note that the same MLV vaccine that induced lower IFN responses in neonatal calves also failed to significantly increase IFN levels above baseline levels when analyzing nasal secretions collected from pregnant and post-partum dairy cows (Cortese et al. 2017). A remarkable observation in this study was high concentrations of both Type I and II IFNs were present in nasal secretions prior to vaccination. Further study is required to clarify whether this observation is unique to pregnant dairy cows or reflects the sensitivity and specificity of the assays used to quantify IFN. Differences in IFN responses have also been observed following IN vaccination of 6 month old calves with MLV vaccines, containing BHV-1 and PI-3 components. Primary IN vaccination four days prior to BHV-1 challenge significantly increased IFN production, relative to naïve controls, at two days after BHV-1 challenge (Hill et al. 2019). The levels of IFN secretion induced by the IN vaccine used in this study were similar to those induced by challenge with a virulent BHV-1 isolate. In contrast, a secondary IN vaccination given 4 days prior to BHV-1 challenge significantly reduced IFNα and IFNγ production following BHV-1 challenge. The reduction in IFN production in vaccinated versus unvaccinated calves may be explained by a marked reduction in BHV-1 infection in IN vaccinated calves since BHV-1 is a potent inducer of IFN secretion in the URT (Osman et al. 2017). Regulation of the innate IFN response by mucosal vaccine-induced effector or memory cells was, however, suggested by the authors (Hill et al. 2019). An immune regulatory mechanism was postulated since there was a similar reduction in IFN secretion in calves receiving a primary IN vaccination as neonates and SC booster with the same vaccine 4 days prior to BHV-1 challenge. Calves receiving the SC booster vaccination failed to develop an IFN response despite shedding of BHV-1 in nasal secretions at levels similar to the naïve control group. This observation indicated IFN responses were not determined solely by the level of BHV-1 replication in the URT. Regulation of pro-inflammatory innate immune responses, such as IFNγ production, may be an important mechanism by which mucosal vaccination could be used to minimize inflammation and damage at mucosal surfaces. There is increasing evidence that acquired immunity can regulate innate immune responses (Okeke and Uzonna, 2019) suggesting this aspect of mucosal vaccines should be further investigated.

Mucosal Vaccines for Bacterial Respiratory Pathogens

The BRD complex is characterized by a broad range of bacterial pathogens that are isolated from lungs of animals with fatal pneumonia. Many of these bacterial pathogens are opportunistic pathogens, residing in the URT of healthy animals. Analysis of the URT microbiome in healthy neonatal calves confirmed bacterial pathogens, such as *Mannhemia*, *Mycoplasma, Moraxella* and Pasteurella colonize the nasopharynx during the first few days of life and persist as residents within the URT microbiome (Lima et al. 2016). It was not known, however, if these commensal bacteria induced local mucosal immune responses. Midla et al (In Press) recently reported healthy neonatal calves develop IgA antibody responses to both *Mannheimia haemolytica* and *Pasteurella multocida*. An IgA response was detected in nasal secretions within 3 to 7 days after birth and IgA titres gradually increased over the first 6 weeks of life. These results are consistent with early colonization of the URT (Lima et al. 2016) and provide evidence the mucosal immune system samples and responds to potential respiratory pathogens. Studies are now needed to determine whether these local IgA responses influence URT colonization by opportunistic pathogens. Bacterial abundance in the URT of neonatal calves increases during the first 6 weeks of life and opportunistic pathogens, such as Mannheimia, Mycoplasma, and Pasteurella, remain among the most abundant genera (Lima et al. 2016). Therefore, there does not appear to be an inverse relationship between bacterial abundance and developing IgA antibody responses.

Observations on URT microbiome (Lima et al. 2016) and IgA responses to *Mannheimia haemolytica* and *Pasteurella multocida* (Midla et al. In Press) raise questions regarding the efficacy of IN vaccines designed to enhance IgA production in the URT and whether immune responses in the URT can prevent bacterial infection in the lung. This question has become even more important since Midla et al. (In Press) reported that IN vaccination of newborn calves with MLV vaccines significantly altered IgA responses to *Pasteurella multocida*. Further, Ayalew et al. (2009) also reported high levels of IgA antibody reacting with *Mannheimia haemolytica* whole cells and leukotoxin in weaned beef calves prior to IN vaccination. This observation indicates IgA responses to *Mannhaimia haemolytica* that develop in nenonatal calves persist for at least the first 6 months of life. Furthermore, IN vaccination of weaned calves with a recombinant protein, containing an immunodominant epitope of the *Mannheimia haemolytica* PlpE protein fused with a neutralizing epitope of leukotoxin covalently linked to truncated cholera toxin subunit B (CTB), induced only small increases in *Mannheimia haemolytica*-specific IgA responses in nasal secretions (Ayalew et al. 2009). IN vaccination did, however, stimulate a significant ($p<0.05$) increase in serum antibody responses to the vaccine epitopes. Vaccination was also associated with a significant reduction in clinical disease following intratracheal challenge with *Mannheimia haemolytica*. These data suggest that the systemic immune response induced by IN vaccination may have contributed to this immune protection against pulmonary infection.

There has also been one report of an oral vaccine containing live-modified *Mannheimia haemolytica* and *Pasteurella multocida* that provided protection against clinical BRD (Briggs and Tatum, 1999). No peer-reviewed studies with the experimental vaccine have been published and the original report did not provide information on either systemic or mucosal immune responses. An oral vaccine that interacts with the enteric mucosal immune system and alters host responses to opportunistic bacterial pathogens in the lung raises questions regarding what immune effector cells may traffic from the intestine to the lung. There is increasing evidence that interactions between the intestinal microbiome and the enteric immune system results in the induction of regulatory T cells, effector lymphocytes, and ILCs that may influence lung inflammation (reviewed in Sokolowska et al. 2018). Further studies are warranted to determine if exposure of the enteric mucosal immune system to *Mannheimia haemolytica* and *Pasteurella multocida* can induce antigen-specific effector cells or ILCs that control pulmonary inflammation. BRD caused by these bacteria is characterized by a profound inflammatory response resulting in extensive tissue damage in the lung (Slocombe et al. 1985; Brogden et al. 1995). An oral vaccine that induced either acquired or innate regulatory cells able to decrease pulmonary inflammation may be an effective strategy to minimize tissue damage when opportunistic pathogens invade the lung. The small intestine is a critical site for the induction of regulatory T cells (Belkaid and Oldenhove, 2008) and induction of Tregs by an oral vaccine would provide a new strategy to control fatal inflammation in the lung.

Mucosal Vaccines Targeting Bovine Enteric Pathogens

MALT in the bovine small intestine is characterized by two functionally and structurally distinct types of Peyer's patches (PP) (reviewed Griebel and Hein, 1996).

These two types of PPs are referred to as discrete and continuous PPs with discrete PPs distributed throughout the jejunum and the continuous PP located in the terminal 2 to 3 m of the small intestine of young calves. Discrete PPs are present at birth, function as immune induction sites for the production of IgA B cells, and persist throughout the life of the animal. In contrast, the continuous PP develops *in utero* but involutes at the time of sexual maturity. The continuous PP functions as an antigen-independent site for diversification of the B cell antibody repertoire (Liljavirta et al. 2013). Recent analyses of the interaction between MAP and MALT in the small intestine of newborn calves confirmed the discrete but not the continuous PPs functions as immune induction sites for IgA antibody responses to an enteric pathogen (Facciuolo et al. 2016). Furthermore, immune responses induced in discrete PPs were able to control MAP infection but the absence of an acquired immune response in the continuous PP was associated with persistent MAP infection (Facciuolo et al. 2020). These observations on PP immune function have important implications when considering the use of oral vaccines in young calves. Uptake of a MLV vaccines or vaccine antigens by the discrete PPs is essential for induction of mucosal effector T and B cells. Whereas, vaccine uptake by the continuous PP would not induce an acquired immune. Natural and inducible B regulatory cells have also been identified in ruminant PPs (Jimbo et al. 2016; Jimbo et al. Manuscript Submitted). It is not known, however, if these regulatory B cells influence the induction of acquired immune responses or play a role in oral tolerance. Thus, further studies are required to determine if oral vaccines can induce both mucosal effector cells and regulatory T and B cells.

Viral and bacterial causes of enteritis are primarily a concern in neonatal calves and parenteral vaccination of dams, to enhance passive transfer of specific maternal IgG, is an effective strategy to reduce the risk of neonatal diarrhea (reviewed in Hodgins et al. 2015). Oral and IN vaccines for the enteric viral pathogens, bovine rotavirus and coronavirus, are currently available for immunization of newborn calves (Table 1). Increased IgM and IgA antibody were detected in feces within 5 days after either oral rotavirus vaccination or experimental rotavirus infection of calves that received colostrum lacking rotavirus-specific IgG antibody (Van Zaane et al. 1986). Calves receiving colostrum with maternal IgG specific for rotavirus secreted maternal IgG in feces for 2 to 6 days after birth and oral rotavirus vaccination, when maternal antibody was present, failed to induce detectable IgM or IgA responses in feces (Van

Zaane et al. 1986). In the same study, oral challenge of these vaccinated calves with rotavirus resulted in diarrhea and virus shedding in feces. These studies confirmed maternal antibody interfered with oral vaccination when using a MLV vaccine. Thus, the use of oral ML rotavirus and coronavirus vaccines in newborn calves may be most effective in herds where dams have no prior immunity to the enteric viruses. Alternatively, delaying oral vaccination until 6 days after birth, when calves no longer pass maternal IgG in feces, may improve vaccine immunogenicity and disease protection. The bovine intestinal mucosal immune system changes rapidly during the first week of life (Liang et. 2014) and allowing time for these developmental changes to occur may also increase immune competence and responses to oral immunization.

IN vaccination with a modified-live bovine coronavirus may circumvent the problem of maternal antibody interference in the intestine. It is not known, however, if the enteric disease protection provided by the current IN coronavirus vaccine depends on virus replication in the URT or vaccine virus replication in the small intestine. Another concern is that MLV vaccines for the control of bovine respiratory viruses are frequently given IN to newborn calves. Currently, there are no published studies regarding the effects of co-delivering two or more IN vaccines. The capacity of some MLV vaccines, containing BHV-1 and PI-3, to induce IFN responses in the URT of newborn calves (Midla et al. In Press), raises the possibility that vaccine-induced IFN may inhibit replication of some modified-live vaccine viruses. For example, bovine coronavirus is very sensitive to IFN inhibition of viral replication and bovine coronavirus replication is associated with inhibition of key IFN signaling pathways (Aich et al. 2007). Therefore, if a MLV vaccine, containing BHV-1 and PI-3, is co-administered IN with a bovine coronavirus vaccine then IFN could possibly inhibit bovine coronavirus replication. Preventing or reducing bovine coronavirus replication may inhibit vaccine-induced immune responses. Alternatively, the IFN induced by modified-live BHV-1 and PI-3 may be important for the development of protective immune responses to these viruses. If co-delivery of the modified-live bovine coronavirus vaccine inhibits IFN production in the URT then the immunogenicity of the BHV-1 and PI-3 vaccines could be reduced. Another possibility is that bovine coronavirus inhibition of the IFN response may result in increased replication of the modified-live BHV-1 vaccine virus (Osman et al. 2017). The attenuated BHV-1 vaccine virus could then cause excessive tissue damage in either the URT or lung, predisposing calves to secondary bacterial pneumonias (Yates, 1982). As more IN vaccines are developed and become available for use in cattle it will be critical to address issues of vaccine interference and safety before recommending co-administration of two or more IN vaccines.

Mucosal Vaccines Targeting Bovine Ocular Pathogens

The ruminant eye has an abundance of isolated and aggregated lymphoid follicles in the conjunctiva and third eyelid but lacks typical lymphoepithelial structures associated with organized lymphoid tissue (Chodosh et al., 1998). An abundance of dendritic cells (DCs) in both the bovine cornea (Bryan et al., 1988) and conjunctiva (Mirabzadeh-Ardakani et. al., 2016) suggests ample capacity to sample local antigens and IgA is the predominant Ig isotype present in tears. Thus, the conjunctiva is considered a potential site for delivery of mucosal vaccines in ruminants (Pugh et al. 1985; Sanchis et al., 1995) and immune induction sites may involve both the conjunctiva associated lymphoid tissue (CALT) and draining cephalic lymph nodes. The bovine eye has the capacity to respond to both infectious and non-infectious stimuli with CALT and IgA providing a protective barrier that destroys pathogens without causing an inflammatory response that could destroy the integrity of the eye (Knop and Knop 2007).

A variety of viral and bacterial pathogens are associated with infectious ocular disease in cattle but *Moraxella* (*M.*) *bovis* is the principal bacterial agent associated with the ocular disease known as infectious bovine keratoconjunctivitis (IBK) (reviewed in Baptista, 1979; Angelos, 2015). Field studies have shown, however, that *M. bovoculi* is frequently isolated from the eye of calves with IBK, especially if calves were previously vaccinated against M. bovis (Angelos, et al. 2007; O'Connor, et al. 2012). Experimental ocular challenge studies with M. bovoculi by itself have failed to induce corneal edema and ulcers (Angelos, 2010). At this time, it can only be stated that since *M. bovoculi* also possesses cytotoxin activity and pili that it may assist M. bovis in causing IBK (Angelos 2015; Dickey, et al. 2018).

IBK was first described decades ago but a vaccine that effectively prevents *Moraxella* ocular infections is not available. Pilin-based vaccines containing multiple M. bovis strains are the most common commercial vaccines in use since the 1980's. These vaccines are injected parenterally and have been reported to reduce the incidence and severity of IBK infections (Lepper and Hermans 1986, Angelos, et al. 2007). However, vaccine failure has been reported for randomized blinded control trials with both M. bovis (Angelos et al. 2007; Cullen et al. 2017a) and *M. bovoculi* (Angelos et al. 2010; O'Connor et al. 2019) bacterins and recombinant protein vaccines. Serum antibody titres were not monitored during field trials with commercial vaccines so the immunogenicity of commercial vaccines was not confirmed. The trial by Angelos et al. (2007) included serology and increased serum antibody titres were observed following vaccination but increased IgG antibody did not correlate with protection. Therefore, induction of a systemic immune response with increased serum IgG antibody titres may not provide a good correlate of immune protection.

Failure of parenteral vaccines to prevent IBK has been attributed to several possible factors. These factors include vaccine failure to cross-protect against the many different strains of M. bovis and M. bovoculi (Dickey et al. 2018), low vaccine immunogenicity after one or two vaccinations, and failure of IgG antibody to be secreted in ocular secretions. Furthermore, there is evidence that *Moraxella* spp are members of the commensal bacterial community in both the eye (Cullen et al. 2017b) and URT (Lima et al. 2016). Moraxella spp persistence at mucosal surfaces suggests local IgA antibody responses may provide some protection against colonization and viulence factors, such as pili and hemolysin. IN vaccination of calves did not significantly increase serum IgG titres but significantly increased IgA titres in tears specific for the vaccine antigen (Angelos et al. 2016). Disease protection was not evaluated in this study. Zbrun et al. (2012) also evaluated an IN *M. bovis* pili vaccine and observed increased IgA titres in tears but no significant reduction in M. bovis isolation and IBK lesions when comparing vaccinated and non-vaccinated calves. Evaluation of these experimental IN vaccines was performed with small numbers of animals and relatively small increases in IgA titres were observed in tears. Therefore, it is difficult to assess how effective IN vaccination may be for induction of protective IgA responses in the eye.

The capacity of CALT to respond to commensal *Moraxella* and possibly produce protective levels of IgA antibody production is presently not known. Further studies are required to determine the onset and magnitude of IgA antibody responses to *Moraxella* spp in the tears of both young calves and mature animals. It is critical to understand if alternative mucosal vaccination strategies, possibly directly targeting the conjunctiva (Pugh et al. 1985), can induce more robust and protective immune responses in the eye. Local IgA antibodies may be more effective than serum IgG in blocking bacterial adherence and corneal damage. IgA antibody doesn't fix complement and this may be important for minimizing ocular edema and inflammation when bacterial virulence factors are secreted to initiate onset of IBK. Vaccine safety would be critical, however, since a mucosal vaccine delivered to the eye must minimize any risk of ocular inflammation that would compromise ocular function. Thus, vaccine formulations or delivery vehicles that interact with CALT without causing ocular inflammation would need to be developed.

Mucosal Vaccines Targeting Genital Tract Infections

Important reproductive diseases affect cattle globally and several vaccines have been developed for both protozoan (*Tritrichomonas foetus*) and bacterial (*Leptospira* spp; *Campylobacter fetus*) causes of gential tract infection (reviewed in Hodgins et al. 2005). These vaccines are injected parenterally and no mucosal vaccines are available for genital tract infections. The efficacy of these parenteral vaccines reveals that systemic immunity, including IgG antibody, provides protective immunity (reviewed in Hodgins et al. 2005). Few experimental vaccines have been evaluated, to assess the efficacy of delivering vaccines to bovine vaginal or uterine epithelial surfaces. A DNA vaccine, encoding a BHV-1 gene, was delivered to the vaginal epithelium and induced specific IgG and IgA antibody responses in blood and nasal secretions (Loehr et al. 2001). In this study, immune responses also occurred in the URT and were associated with a significant reduction in clinical disease following aerosol challenge with BHV-1. This study confirmed antigen uptake by the vaginal epithelium and the induction of both IgG and IgA antibody responses. A link between mucosal effector cells induced in the reproductive tract and trafficking of these cells to the URT was also demonstrated. Homing of mucosal effector cells to the genital tract has also been observed following IN immunization of mice (Klavinskis et al. 1999), suggesting shared homing mechanisms for mucosal effector cells induced within the URT and reproductive tract. Further studies are required in cattle to determine if IN vaccination may be an effective route to enhance protection against pathogens invading the reproductive tract.

Alternative Mucosal Vaccine Delivery Vehicles

The importance of mucosal immunity in preventing pathogen invasion has been an important factor driving continued interest in developing mucosal vaccine delivery vehicles for use in cattle. A number of other important considerations make mucosal delivery of vaccines an attractive option for immunizing cattle. These considerations include concerns regarding local tissue reactions at parenteral vaccine injection sites and the impact of these reactions and needle fragments on meat quality (Cresswell et al. 2017). In countries where bovine leukemia virus (BLV) has not been eliminated the injection of vaccines and other veterinary therapeutics contributes to the spread of this disease when the same hypodermic needle is used to inject more than one animal (Bartlett et al. 2014). Rapid development of the mucosal immune system in neonatal calves also provides an opportunity to induce robust immune responses prior to maturation of the systemic immune system and mucosal immunization can circumvent vaccine interference by maternal antibody (Hill et al. 2012). Finally, as the number of cattle in individual beef and dairy herds increases, time and labour costs associated with restraining individual animals for vaccination may reduce the use of vaccines. Thus, developing vaccines that could be delivered through feed or water is an attractive option to eliminate handling of individual animals. Plant-based vaccines expressing *Mannheimia haemolytica* proteins were immunogenic in rabbits (Lee et al. 2001; Lee et al. 2008) but these vaccines have not reported to be effective in cattle.

A number of mucosal delivery vehicles have been tested for use in cattle and these are discussed in terms of the potential advantages or disadvantages they may provide (Table 3). One of the major challenges for mucosal vaccine development is a lack of effective mucosal adjuvants (Newsted et al. 2015). While this is not an issue with MLV vaccines, several strategies to overcome this limitation have been evaluated in cattle (Table 3). Cholera toxin (CT) is a potent stimulant of IgA responses to a wide variety of potential vaccine antigens but is associated with adverse effects in both the URT and small intestine (Liang et al. 1989). Toxicity has been overcome through the use of CT subunits and creation of fusion proteins consisting of a CT subunit and a variety of vaccine epitopes. Two vaccine epitopes from *Mannheimia haemolytica* were included in a fusion protein with a CT subunit (Ayalew et al. 2009). Following two IN immunizations of weaned beef calves with this chimeric protein there was not a significant increase in IgA production specific for either leukotoxin or whole bacterial cells. These data did not support the conclusion that the CT subunit was an effective mucosal adjuvant. It should be noted, however, calves in the study had high IgA titres to *Mannhimia haemolytica* leukotoxin and whole cells prior to IN immunization. Thus, the vaccine was given to animals that had already responded to *Mannheimia haemolytica* residing as a commensal in the URT.

The need for appropriate mucosal adjuvants was also evident in a study using oil emulsion and alginate hydrogel to deliver recombinant proteins as a vaccine. These adjuvants were effective when the *Staphylococcus aureus* recombinant proteins were delivered SC but not when vaccine was delivered IN or injected into the mammary (Boerhout et al. 2015). The IN delivery of recombinant protein vaccines can, however, induce both local IgA responses and protective immunity when appropriately formulated. McGill et al (2018) formulated BRSV recombinant G and F proteins in polyanhidride nanoparticles and neonatal calves received a single IN immunization. The vaccine resulted in a significant IgA response in both nasal secretions and lung lavage fluid following BRSV challenge and a significant reduction in lung pathology. Limitations with this type of vaccine include the production and purification of recombinant proteins and the relatively low incorporation of approximately 20% of the protein into nanoparticles. The nanoparticles did, however, provide sustained release of protein over a 5-week period. The complexity of vaccine formulation with nanoparticles or larger alginate microspheres (Kim et al. 2002) as mucosal delivery vehicles also represent barriers to commercial vaccine production. Scaling vaccine production to provide millions of vaccine doses requires consistent production protocols that ensure vaccine potency and stability.

The use of ML viral and bacterial vaccines can eliminate the need for mucosal adjuvants, especially if the viral component replicates within mucosal epithelial cells. Replication of the ML vaccine virus or bacteria amplifies the amount vaccine antigen delivered and can prolong the duration of antigenic stimulation. Attenuation of ML pathogens reduces the risk of adverse effects but for some pathogens, such as BHV-1, this does not eliminate persistent or latent infections following IN vaccination (Lemaire et al. 2000). It is possible, however, to engineer the BHV-1 virus, through deletion of the thymidine kinase (tk) gene, to prevent establishment of latent infections (Zhang et al. 2011). A single IN immunization with the glycoprotein G and tk gene deleted virus induced protective immunity that reduced virus shedding and lung pathology when calves were challenged with wild-type BHV-1 28 days after vaccination. Zhang et al. (2011) did not analyze IgA antibody responses in the URT following vaccination so mucosal immune responses to the double gene deleted virus were not characterized. Studies to determine duration of immunity were also not performed so it is not known how important latent infections are for maintaining immune memory following IN vaccination with current BHV-1 containing MLV vaccines (Hill et al. 2019).

Studies were also performed to determine if the immunogenicity of ML respiratory virus vaccines could be improved through insertion of a bovine cytokine gene. Raggo et al. (2000) inserted the bovine IFNγ gene into the genome (BHV-1/IFNγ) of a gC deleted (gC^-) Cooper strain BHV-1 and expression of IFNγ protein in BHV-1/IFNγ infected cells was confirmed. Seronegative, 6-month old calves were injected IN with either wild-type, gC^-, or BHV-1/IFNγ Cooper strains. No significant differences in virus shedding were observed when comparing the gC^- or BHV-1/IFNγ Cooper strains following primary infection or reactivation of latent virus. There was also no difference in IFNγ levels in nasal secretions following primary viral infection and serum IgG antibody responses were similar. The only significant difference observed was lower IgA antibody titres in nasal secretions of the BHV-1/IFNγ infected calves following reactivation of latent virus. This may have been due to an immune modulatory effect of IFNγ since there were similar levels of virus shedding in nasal secretions. Limited immune modulatory effects by the IFNγ transgene may not be surprising when wild-type and gC^- BHV-1 strains induced high levels of IFNγ secretion. A more pronounced immune modulatory effect may occur if the BHV-1/IFNγ virus were evaluated in newborn calves. Midla et al. (In Press) observed IN vaccination of neonatal calves with two different ML BHV-1 vaccines induced low levels of IFNγ in nasal secretions and not all calves produced detectable levels of IFNγ. Insertion of cytokine or other immune regulatory genes into vaccine viruses may enhance vaccine immunogenicity and immune protection if they were used strategically to augment innate or acquired mucosal immune responses when these responses were limited by age or other developmental and environmental factors.

No viral vaccine vectors are currently licensed for mucosal immunization in cattle but several vectors have been evaluated experimentally for immunogenicity and disease protection following IN immunization. Bovine adenovirus-3 (BAdV-3), which is not pathogenic in cattle, has been evaluated with a number of transgenes, including BHV-1 glycoprotein D, BVDV envelope 2, and BCV hemagglutinin-esterase glycoprotein (reviewed in Ayalew et al. 2015). Studies in cattle confirmed IN delivery of the BAdV-3 vaccine vector circumvented vaccine interference by serum antibody induced by natural exposure to wild-type BAdV-3. Further, as a strategy to enhance mucosal IgA production, BAdV-3 was constructed to express both a vaccine antigen and bovine IL-6 but this did not increase IgA production or disease protection (Kumar et al. 2014). With a greater understanding of the bovine mucosal immune system, it may be possible to construct vaccine vectors that enhance dendritic cell (DC) recruitment and activation at the site of vaccine transgene expression. This could further enhance antigen presentation and the induction of mucosal immune responses. Alternatively, it may be possible to engineer vaccine vectors to enhance targeting to specific mucosal immune induction sites to enhance efficiency of vector delivery (reviewed in Ayalew et al. 2015). Bovine viral pathogens, such as BHV-1 (Kühnle et al. 1998) and BHV-4 (Williams et al. 2019), have also been evaluated as potential mucosal vaccine vectors. These vectors have a natural tropism for mucosal epithelium and the potential advantage of requiring a single immunization. Establishment of latent infections by these herpesviruses and subsequent reactivation of latent infections can boost immune responses (Williams et al. 2019).

Most alternative mucosal delivery systems have focused on IN delivery as the primary route of vaccine delivery. A major disadvantage of IN vaccine delivery is the need to restrain the head of animals for vaccine delivery. Oral vaccine delivery creates the possibility of vaccine delivery through either feed or water, eliminating the requirement to restrain individual animals. This would greatly reduce labour costs and time associated with vaccine delivery. It may also be possible to deliver vaccines to free ranging animals prior to high risk periods of infection, such as weaning suckling beef calves and subsequent congregation of these naïve animals in feedyards. Oral vaccine delivery in feed or water must, however, address critical issues such as vaccine formulation and stability, efficiency of vaccine delivery, and ensuring consistent vaccine dosing for all animals. Vaccine delivery will require vaccines that survive passage through the rumen and vaccines that are efficiently sampled by MALT in the small intestine. Thus, few oral vaccines or oral vaccine delivery systems have been evaluated in cattle (Table 3).

González Cano et al. (2017) evaluated bacteriophage as a potential oral vaccine delivery vehicle in neonatal calves. Bacteriophage provide a potentially effective vaccine delivery vehicle since they are resistant to low pH in the stomach and González Cano et al. (2017) confirmed lambda phage (LP) survived passage through the gastro-intestinal tract of mice. A LP display particle (LDP) was constructed displaying three known B-cell epitopes and experiments performed to confirm LP uptake by MALT and vaccine immunogenicity in the small intestine of newborn calves. B cell epitope-specific IgA antibody responses were detected in the small intestine MALT following a single enteric immunization with LDP without a mucosal adjuvant. These experiments established proof-of-principle that LDP are an effective oral vaccine delivery vehicle in newborn calves. Production of LDP in bacterial cells means this vaccine delivery system is scalable for production of a commercial vaccine. Major limitations of this vaccine delivery system are the size of protein epitopes that can be efficiently displayed on the surface of LP particles and a requirement that protective B or T cell epitopes be identified for display. Gamage et al. (2009) demonstrated that both T and B cell epitopes can be displayed on the same LDP and both T and B cell effector responses were induced.

There is only a single report of genetically-modified bacteria being used as an oral vaccine in cattle. Briggs and Tatum (1999) reported that gene deleted *Mannheimia haemolytica* and *Pasteurella multocida* could be sprayed on feed eaten by weaned calves and oral vaccination was associated with a significant reduction in BRD. No further studies have been reported with this vaccine and neither systemic or mucosal immune responses were characterized in this study. Thus, it is unknown if oral vaccination altered immune responses to these opportunistic pathogens in the lung or URT or altered systemic immune responses. Also, it is not known if the vaccine increased antibody or effector T cell responses or possibly induced regulatory T cells following oral exposure.

Future Challenges

Over the last 10 years it has been accepted that IN vaccines provide an effective strategy to avoid antibody interference and establish protective immunity in neonatal calves to respiratory viral pathogens (Hill et al. 2012). Vaccines that prevent bacterial infections of the lung, eye, gastro-intestinal tract, and mammary gland have proven

Table 3. Experimental Mucosal Delivery Vehicles Evaluated in Cattle

Delivery Vehicle	Vaccine Target	Delivery Route	Advantage	Limitation	Reference
Alginate hydrogel	S. aureus	IN Intramammary	Non-invasive	Low immunogenicity	Bouerhout et al. 2015
CTB Fusion	*Mannheimia haemolytica*	IN	Non-invasive	Low immunogenicity	Ayalew et al. 2009
Gene-deleted Virus	BHV-1	IN	No latent infection	Unknown DOI[2]	Zhang et al. 2011
Gene-Inserted Virus	BHV-1	IN	Unchanged virulence	No gain in function	Raggo et al. 2000
Engineered Bacteria	*Mannheimia haemolytica*	Oral			Briggs, Tatum. 1999
Lambda Phage	BSE[3]	Oral			González Cano et al. 2017
Nanovaccine[1]	BRSV	IN	Single Vaccination	Complex formulation	McGill et al. 2018
Viral Vector	BHV-1; BVDV; BCV	IN	Circumvent pre-exiting immunity	Requires high vector dose	Ayalew et al. 2015; Williams et al. 2019

[1]Polyanhydride nanoparticles
[2]DOI: Duration of Immunity
[3]Bovine spongiform encephalopathy disease associated prion protein

more difficult to develop (Table 4). Each of these pathogens have unique properties but share some similarities that may contribute to the on-going challenge in developing effective mucosal vaccines. Many of the bacterial pathogens causing ocular and lung disease reside as commensals in the URT (Lima et al. 2016; Cullen et al. 2017b). These opportunistic pathogens are part of disease complexes characterized by viral co-infections or other environmental factors that compromise the integrity of mucosal defences at the site of infection. Studies now confirm that these bacteria are recognized by the mucosal immune system in the URT shortly after colonization occurs in neonatal calves (Midla et al. In Press). This information raises the question whether IN delivery of vaccines can provide additional benefit in the control of these opportunistic pathogens. Studies are required to determine whether IgA production in the URT has a direct impact on the abundance of these opportunistic pathogens. Data from Lima et al (2016) indicates the abundance of opportunistic pathogens, such as *Mannheimia* and *Pasteurella*, increases in the URT of neonatal calves during the first 6 weeks of life. This increase in bacterial abundance coincides with increasing IgA titres specific to these bacteria (Midla et al. In Press). It remains to be determined, however, whether altering the repertoire of this IgA response to react more with specific bacterial attachment proteins could change bacterial abundance. It also remains to be determined whether the overall abundance of opportunistic pathogens, or even specific variants of the pathogens, in the URT is a significant factor contributing to colonization of either the lung (Amat et al. 2020) or the eye.

An alternative approach to the development of mucosal vaccines that ameliorate the effects of these opportunistic pathogens is to consider vaccine development that prevents the inflammatory responses that cause life-threatening damage at mucosal surfaces. *Mannheimia* and *Pasteurella* pneumonia are diseases characterized by profound inflammation and destruction of lung tissue and IBK is associated with destructive inflammation in the conjunctiva and cornea. These inflammatory responses to bacterial pathogens compromise tissue function and, in the lungs, can be life threatening. Previous vaccines focused on induction of effector responses, either antibody or T cells, that control or clear bacterial infection. These effector responses may, however, contribute to the immune pathology of disease and disruption of mucosal tissue function. Therefore, an alternative mucosal vaccine approach may be to focus on the induction of innate or acquired immune regulatory cells that prevent tissue damage. The small intestine is a major site for the induction of both regulatory T and B cells and the induction of IL-22 secreting ILCs. This raises the possibility that oral vaccines, if appropriately formulated and designed, may provide a strategy to induce regulatory cells controlling inflammation in either the lung or eye. Reagents and assays are now available to identify bovine regulatory T and B cells, making it possible to detect these responses if they develop following oral vaccination.

Another group of challenging bacterial pathogens are those that establish persistent infections at mucosal sites (Table 4). Persistent *Staphylococcus aureus* (Rainard et al. 2019) and *Streptococcus uberis* (Leigh, 1999) of the bovine mammary gland and persistent MAP infection of the intestine (Barkema et al. 2017) are global challenges to bovine health. The bovine mammary gland shares many immune functions with the systemic or peripheral immune system (reviewed in Kehrli and Harp, 2001) and vaccines developed to control mammary gland pathogens have focused on parenteral vaccine delivery. Control of inflammation in the mammary gland is critical for sustaining gland function and ensuring milk quality. Possible linkages between bovine mucosal immunity and homing of mucosal effector or regulatory cells to the mammary gland have not been studied (Sordillo, 2018). Thus, there is little available knowledge to suggest mucosal vaccines may provide an alternative or better strategy to enhance immune protection in the mammary gland.

Control of enteric MAP infection presents a global problem, not only for cattle but also for small ruminants and camels. Furthermore, control of MAP infections is compounded by a lack of effective diagnostic tools to detect infected animals, a long period of subclinical infection characterized by intermittent shedding of MAP in feces, and persistence of MAP in the environment (Barkema et al. 2017). There are also unique aspects of the MAP-host interaction that further complicate development of a vaccine to control MAP infection and reduce fecal shedding. MAP is preferentially taken up by M-cells located within the follicle associate epithelium overlying GALT and recent studies have confirmed that MAP is taken up equally well by both the discrete and continuous PPs located within the small intestine of neonatal calves (Facciuolo et al. 2020). There are, however, very different host responses to MAP following infection of discrete versus continuous PPs. The discrete PPs function as mucosal immune induction sites and within one to two months post-infection IgA antibody responses can be detected (Facciuolo et al. 2016) and MAP infection is eventually controlled (Facciuolo et al. 2020). In contrast, the continuous PP is a primary lymphoid organ and MAP can persist for months within B lymphoid follicles without inducing an IgA response. Thus, MAP infects a primary lymphoid tissue in the small intestine where it remains undetected by the host immune system (Liang et al. 2016). This creates a challenge for MAP vaccine development since vaccine-induced induce effector cells must home to a primary B lymphoid tissue. Another challenge for vaccine development is that recent studies confirmed extensive bacterial colonization of the bovine gastrointestinal tract during birth and prior to parturition (Malmuthuge et al. 2019). Calves born to cows infected with MAP may be infected either during birth or through environmental MAP exposure during the neonatal period. Thus, a vaccine to prevent MAP infection would need to be effective when given to newborn calves and rapid onset of protective immunity would be necessary. Alternatively, a therapeutic vaccine would be required that is effective when given after MAP infection is established.

Table 4. Mucosal Bacterial Pathogens

Mucosal Site	Pathogen	Commensal	Persistent Pathogen
Eye	*Moraxella*	√	
Lung	*Mannheimia,*	√	
	Pasteurella,	√	
	Mycoplasma	√	
Intestine	MAP[1]		√
Mammary	*Staphylococcus*		√
Gland	*Streptococcus*		√

[1]MAP: *Mycobacterium avium* subspecies *paratuberculosis*

An experimental modified-live MAP vaccine was delivered orally to goats but failed to provide significant protection against enteric infection (Hines et al. 2015). Efficacy of an oral vaccine may be limited since vaccine uptake by the continuous PP will fail to induce an immune response (Facciuolo et al. 2016). Further, if the vaccine induces an immune response in discrete PPs it is not known if mucosal effector cells generated in this GALT will subsequently home to the continuous PP. There is, however, a report that a MAP bacterin, injected parenterally, can reduce clinical disease and fecal shedding of bacteria by MAP infected cattle (Tewari et al. 2014). A recent study confirmed the same vaccine given to neonatal calves could significantly reduce MAP burden in the continuous PP and MAP shedding into the intestinal lumen (Facciuolo et al. Manuscript Submitted). The parenteral MAP vaccine was given two months prior to bacterial challenge and a significant reduction in bacterial burden was observed one month after infection. This study provides remarkable evidence that a parenteral vaccine induced effector cells that homed to the continuous but not the discrete PPs. Thus, parenteral vaccination may provide a strategy to induce effector cells that can respond to MAP infection at the most vulnerable site within the small intestine. Further studies are required to determine if this phenomenon is unique to this particular mycobacterial vaccine. Co-formulating recombinant BHV-1 gD protein with bovine IL-1 enhanced IgA antibody responses in the URT following subcutaneous vaccine injection (Gao et al. 1995). In this study, the BHV-1 vaccine was targeted to a lymph node that drains the URT so there was a functional link between the parenteral vaccine injection site and the mucosal surface of the URT. In contrast, the parenteral MAP bacterin was injected SC on the neck of neonatal calves. The MAP bacterin may provide information that can be used to understand how systemic immunity can be linked to mucosal immunity in the small intestine. This knowledge could guide development of vaccines that effectively control MAP infection and shedding from at least one region of the intestinal tract.

Conclusions

Mucosal vaccines have become an integral part of bovine health management, especially for the control of respiratory and enteric viral infections in young calves. Efficacy of these vaccines arises from an avoidance of maternal antibody interference, rapid development of MALT in newborn calves, and rapid onset of protective immunity following vaccination. These attributes make mucosal vaccines an effective tool for the control of acute viral infections with a tropism for mucosal epithelium. Substantial challenges remain, however, in developing effective mucosal vaccines targeting pathogens residing with the URT microbiome, especially since these opportunistic pathogens induce IgA antibody responses early in life. Further work is required to identify the immune responses induced by a mucosal vaccine that can enhance control of opportunistic pathogens but minimize the inflammation and immune pathology that disrupts vital mucosal tissue functions. Success in this area may require a much better understanding of how mucosal vaccines induce innate or acquired immune regulatory cells.

Persistent bacterial pathogens in the mammary gland and the gastro-intestinal tract also present future challenge for mucosal vaccines. Regionality of the mucosal immune system with restricted homing of mucosal effector cells may require unique vaccination strategies to control infections at individual mucosal sites. A single vaccination strategy may not be sufficient to control MAP infection throughout the entire intestinal tract and two or more vaccine modalities, such as a combination of systemic and mucosal vaccines, may be required. Furthermore, a combination of prophylactic and therapeutic vaccines may also be required when MAP infection occur during or soon after birth. The success of future mucosal vaccines will also depend on developing vaccine delivery systems and adjuvants that target vaccines to appropriate mucosal sites and ensure the induction of effector or regulatory cells that control infection without disruptingf the mucosal barrier and tissue function.

Acknowledgements

We wish to thank Juliane Deubner for her design and preparation of figures. Dr Philip Griebel is funded by a Tier I CRC in Neonatal Mucosal Immunology provided by Canada Institutes for Health Research (CIHR).

References

Aich, P., Wilson, H.L., Kaushik, R.S., Potter, A.A., Babiuk, L.A., Griebel, P. (2007). Comparative analysis of innate immune responses following infection of newborn calves with bovine rotavirus and bovine coronavirus. J Gen Virol 88(Pt 10), 2749-61.

Amat, S., Alexander, T.W., Holman,D.B., Schwinghamer,. T., Timsit, E. (2020). Intranasal Bacterial Therapeutics Reduce Colonization by the Respiratory Pathogen Mannheimia haemolytica in Dairy Calves. mSystems 5(2), e00629-19.

Angelos, J.A. (2010). Moraxella bovoculi and infectious bovine keratoconjunctivitis: cause or coincidence? Vet Clin North Am Food Anim Pract 26(1), 73-78.

Angelos, J.A. (2015). Infectious bovine keratoconjunctivitis (pinkeye). Vet Clin North Am Food Anim Pract 31(1), 61-79.

Angelos, J.A., Bonifacio, R.G., Ball, L.M., Hess, J.F. (2007). Prevention of naturally occurring infectious bovine keratoconjunctivitis with a recombinant Moraxella bovis pilin-Moraxella bovis cytotoxin-ISCOM matrix adjuvanted vaccine. Vet Microbiol 125(3-4), 274-83.

Angelos, J.A., Chigerwe, M., Edman, J.M., Hess, J.F. (2016). Systemic and ocular immune responses in cattle following intranasal vaccination with precipitated or partially solubilized recombinant Moraxella bovis cytotoxin adjuvanted with polyacrylic acid. Am J Vet Res 77(12), 1411-1418.

Aubry, P., Warnick, L.D., Guard, C.L., Hill, B.W., Witt, M.F. (2001). Health and performance of young dairy calves vaccinated with a modified-live Mannheimia haemolytica and Pasteurella multocida vaccine. J Am Vet Med Assoc 219, 1739–42.

Ayalew, E., Kumar, P., Gaba, A., Makadiya, N., Tikoo, S.K. (2015). Bovine adenovirus-3 as a vaccine delivery vehicle. Vaccine 33(4), 493-99.

Ayalew, S., Step, D.L., Montelongo, M., Confer, A.W. (2009). Intranasal vaccination of calves with Mannheimia haemolytica chimeric protein containing the major surface epitope of outer membrane lipoprotein PlpE, the neutralizing epitope of leukotoxin, and cholera toxin subunit B. Vet Immunol Immunopathol 132(2-4), 295-302.

Baptista, P.J. (1979). Infectious bovine keratoconjunctivitis: a review. Br Vet J 135(3), 225-42.

Barkema, H.W., Orsel, K., Nielsen, S.S., Koets, A .P., Rutten. V.P., Bannantine, J,P,, Keefe, G.P., Kelton, D.F., Wells, S.J., Whittington, R.J., Mackintosh, C.G., Manning, E.J., Webe,r M.F., Heuer, C., Forde, T.L., Ritter, C., Roche, S., Corbett, C.S., Wolf, R., Griebel, P.J., Kastelic, J.P., De Buck, J. (2018). Knowledge Gaps That Hamper Prevention and Control of Mycobacterium Avium Subspecies Paratuberculosis Infection. Transbound Emerg Dis 65 Suppl 1, 125-148.

Bartlett, P.C., Sordillo, L.M., Byrem, T.M., Norby, B., Grooms, D.L., Swenson, C.L., Zalucha, J., Erskine, R.J. (2014). Options for the control of bovine leukemia virus in dairy cattle. JAVMA 244(8), 914-22.

Belkaid, Y., Oldenhove, G. (2008). Tuning Microenvironments: Induction of Regulatory T Cells by Dendritic Cells, Immunity 29(3), 362–371.

Belkaid, Y., Tarbell, K. (2009). Regulatory T cells in the control of host-micro-organism interactions. Annu Rev Immunol 27, 551-89.

Berod L, Puttur F, Huehn J, Sparwasser T. Tregs in infection and vaccinology: heroes or traitors? Microb Biotechnol. 2012, 5(2): 260–269.

Bielefeldt Ohman, H., Babiuk, L.A. (1986). Alteration of some leukocyte functions following in vivo and in vitro exposure to recombinant bovine alpha- and gamma-interferon. J Interferon Res 6(2),123-136.

Boerhout, E., Vrieling, M., Benedictus, L., Daemen, I., Ravesloot, L., Rutten, V., Nuijten, P., van Strijp, J., Koets, A., Eisenberg, S. (2015). Immunization routes in cattle impact the levels and neutralizing capacity of antibodies induced against S. aureus immune evasion proteins. Vet Res 46,115.

Briggs, R.E., Tatum, F.M. (1999). New mucosal vaccine in beef cattle imparts rapid resistance to pneumonic pasteurellosis after mass-medicating on feed. Presented at the US Animal Health Association Mtg. San Diego, CA.

Brodersen, B.W. (2010). Bovine respiratory syncytial virus. Vet Clin North Am Food Anim Pract 26(2), 323-33.

Brogden, K.A., Ackermann, M.R., Debey, B.M. (1995). Pasteurella haemolytic lipopolysaccharide-associated protein induces pulmonary inflammation after bronchoscopic deposition in calves and sheep. Infect Immun 63, 3595-3599.

Bryan L.A., Griebel, P., Allan, J.R., Davis, W.C. (1988). Identification of bovine epidermal Langerhans cells (LC) using anti-Class II antibody. J. Histochem. Cytochem 36, 991-995.

Burakova, Y., Madera, R., McVey, S., Schlup, J.R., Shi, J. (2018). Adjuvants for Animal Vaccines. Viral Immunol 31(1), 11-22.

Chamorro, M.F., Woolums, A., Walz, P.H. (2016). Vaccination of calves against common respiratory viruses in the face of maternally derived antibodies (IFOMA), AHRR 17(2), 79-84.

Chodosh, J., Nordquist, R.E., Kennedy, R.C. (1998). Comparative anatomy of mammalian conjunctival lymphoid tissue: a putative mucosal immune site. Dev Comp Immunol 22(5-6), 621-30.

Connelley, T.K., Longhi, C., Burrells, A., Degnan, K., Hope, J., Allan, A.J. (2014). Nkp46+ CD3+ Cells: A Novel Nonconventional T Cell Subset in Cattle Exhibiting Both Nk Cell and T Cell Features. J Immunol 192(8), 3868-80.

Cortese, V.S., Grooms, D.L., Ellis, J., Bolin, S.R., Ridpath, J.F., Brock, K.V. (1998). Protection of pregnant cattle and their fetuses against infection with bovine viral diarrhea virus type 1 by use of a modified-live virus vaccine. Am J Vet Res 59(11), 1409-1414.

Cresswell, E., Remnant, J., Butterworth, A., Wapenaar, W. (2017). Injection-site lesion prevalence and potential risk factors in UK beef cattle. Vet Rec 180(3), 70.

Cullen, J.N., Engelken, T.J., Cooper, V., O'Connor, A.M. (2017a). Randomized blinded controlled trial to assess the association between a commercial vaccine against Moraxella bovis and the cumulative incidence of infectious bovine keratoconjunctivitis in beef calves. J Am Vet Med Assoc 251(3), 345-351.

Cullen, J.N., Lithio, A., Seetharam, A.S., Zheng, Y., Li, G., Nettleton, D., O'Connor, A.M. (2017b). Microbial community sequencing analysis of the calf eye microbiota and relationship to infectious bovine keratoconjunctivitis. Vet Microbiol 207, 267-279.

Czarnieckil, C.W., Anderson, K.P., Fennie, E.B., Bielefeldt Ohmann, H., Babiuk, L.A. (1985). Bovine interferon-alpha$_{I}$1 is an effective inhibitor of bovine herpes virus-1 induced respiratory disease. Antiviral Res Suppl 1, 209-215.

Dickey, A.M., Schuller, G., Loy, J.D., Clawson, M.L. (2018). Whole genome sequencing of *Moraxella bovoculi* reveals high genetic diversity and evidence for interspecies recombination at multiple loci. PLoS One 13(12), e0209113.

Duff, G.C., Galyean, M.L. (2007). Board-invited review: recent advances in management of highly stressed, newly received feedlot cattle. J Anim Sci 85, 823-840.

Ellis, J., Gow, S.P., Mahan, S., Leyh, R. (2013). Duration of immunity to experimental infection with bovine respiratory syncytial virus following intranasal vaccination of young passively immune calves. JAVMA 243(11),1602-8.

Ellis, J.A., Gow, S.P., West, K., Waldner, C., Rhodes, C., Mutwiri, G., Rosenberg, H. (2007). Response of calves to challenge exposure with virulent bovine respiratory syncytial virus following intranasal administration of vaccines formulated for parenteral administration. JAVMA 230(2), 233-43.

Ellis, J., West, K., Cortese, V., Konoby, C., Weigel, D. (2001). Effect of maternal antibodies on induction and persistence of vaccine-induced immune responses against bovine viral diarrhea virus type II in young calves. JAVMA 219, 351-56.

Endsley, J.J., Ridpath, J.F., Neill, J.D., Sandbulte, M.R., Roth, J.A. (2004). Induction of T lymphocytes specific for bovine viral diarrhea virus in calves with maternal antibody. Viral Immunol 17, 13-23.

Facciuolo, A., Gonzalez-Cano, P., Napper, S., Griebel, P.J., Mutharia, L.M. (2016). Marked Differences in Mucosal Immune Responses Induced in Ileal versus Jejunal Peyer's Patches to Mycobacterium avium subsp. paratuberculosis Secreted Proteins following Targeted Enteric Infection in Young Calves. PLoS One 11(7), e0158747.

Facciuolo, A., Lee, A.H., Gonzalez-Cano, P., Townsend, H.G.G., Falsafi, R., Gerdts, V., Potter, A., Napper, S., Hancock, R.E.W., Mutharia, L.M., Griebel, P.J. (2020). Regional Dichotomy in Enteric Mucosal Immune Responses. to a Persistent Mycobacterium avium ssp. paratuberculosis Infection. Front Immunol 11, 1020.

Fairbanks, K.F., Campbell J., Chase, C.C.L. (2004). Rapid onset of protection against infectious bovine rhinotracheitis virus with a modified-live virus multivalent vaccine. Vet Ther 5, 17-25.

Fries, P.N., Griebel, P.J. (2011). Mucosal dendritic cell diversity in the gastrointestinal tract. Cell Tissue Research 343, 33-41.

Gaertner, F.H., Babiuk, L.A., Mutwiri, G., Armstrong, J.M., Griebel, P.J. (2009). Amended recombinant cells (ARCTM) expressing bovine IFN-γ: An economical and highly effective adjuvant system. Vaccine 27, 1377-1385.

Gamage, L.N., Ellis, J., Hayes, S. (2009). Immunogenicity of bacteriophage lambda particles displaying porcine Circovirus 2 (PCV2) capsid protein epitopes. Vaccine 27, 6595–604.

Gao, Y., Daley, M.J., Splitter, G.A. (1995). BHV-1 glycoprotein 1 and recombinant interleukin 1 beta efficiently elicit mucosal IgA response. Vaccine 13(9), 871-77.

González-Cano, P., Gamage, L.A.N., Marciniuk, K., Hayes, C., Napper, S., Hayes, S., Griebel, P.J. (2017). Lambda Display Phage as a Mucosal Vaccine Delivery Vehicle for Peptide Antigens. Vaccine 35, 7256-7263.

Griebel, P.J., Hein, W.R. (1996). Expanding the role of Peyer's patches in B-cell ontogeny. Immunol Today 17, 30-38.

Griebel, P.J., Hill, K., Stookey, J. (2014). How Stress Alters Immune Responses During Respiratory Infection. AHRR 15(2), 161-165.

Hines, M.E. 2nd, Turnquist, S.E., Ilha, M.R., Rajeev, S., Jones, A.L., Whittington, L., Bannantine, J.P., Barletta, R.G., Grohn, Y.T., Katani, R. (2014). Evaluation of novel oral vaccine candidates and validation of a caprine model of Johne's disease. Front Cell Infect Microbiol 4:26.

Hodgins, D.C., Chattha, K., Vlasova, A., Parreño, V., Corbeil, L.B., Renukaradhya, G.J., Saif, L.S. (2015). Mucosal Veterinary Vaccines: Comparative Vaccinology. Chapter 68, Mucosal Immunology (4th Ed), V 2, 1337-1361.

Hodgson, P., Stookey, J., Popowych, Y., Potter, A., Babiuk, L., Griebel, P.J. (2012). Stress significantly increases mortality following a secondary bacterial respiratory infection. Vet Res 43, 21-29.

Hoerlein, A.B., Marsh, C.L. (1957). Studies on the epizootiology of shipping fever in calves. JAVMA 131, 123-127.

Holmgren, J., Czerkinsky, C. (2005). Mucosal immunity and vaccines. Nature Medicine Suppplement 11(4), S45-S53.

Ioannou, X.P., Gomis, S.M., Karvonen, B., Hecker, R., Babiuk, L.A., van Drunen Littel-van den Hurk, S. (2002). CpG-containing oligodeoxynucleotides, in combination with conventional adjuvants, enhance the magnitude and change the bias of the immune responses to a herpesvirus glycoprotein. Vaccine 21(1-2), 127-37.

Jimbo, S., Griebel, P.J., Mutwiri, G. Natural Regulatory B Cells are Present in Bovine Peyer's Patches. Cell Tissue Res (Manuscript Submitted).

Jimbo, S., Griebel, P.J., Townsend, H., Babiuk, L.A., Mutwiri, G. (2016). Evidence for the existence of regulatory and effector B cell populations in Peyer's patches of sheep. Vet Immunol Immunopathol 174, 26-34.

Johansen, F.-E., Kaetzel, C.S. (2011). Regulation of the polymeric immunoglobulin receptor and IgA transport: new advances in environmental factors that stimulate pIgR expression and its role in mucosal immunity. Mucosal Immunology 4, 598-602.

Kehrli, M.E. Jr, Harp, J.A. (2001). Immunity in the mammary gland. Vet Clin North Am Food Anim Pract 17(3), 495-516.

Kim, B., Bowersock, T., Griebel, P., Kidane, A., Babiuk, L.A., Sanchez, M., Attah-Poku, S., Kaushik, R.S., Mutwiri, G.K. (2002). Mucosal immune responses following oral immunization with rotavirus antigens encapsulated in alginate microspheres. J. Controlled Release 85,191-202.

Klavinskis, L.S., Barnfield, C., Gao, L., Parker, S. (1999). Intranasal immunization with plasmid DNA-lipid complexes elicits mucosal immunity in the female genital and rectal tracts. J Immunol 162, 254-26.

Knop, E., Knop, N. (2007). Anatomy and immunology of the ocular surface. Chem Immunol Allergy 92, 36-49.

Kolb, E.A., Buterbaugh, R.E., Rinehart, C.L., Ensley, D., Perry, G.A., Abdelsalam, K.W., Chase, C.C.L. (2020). Potection against bovine repiratory virus in calves vaccinated with adjuvanted modified live vaccine administered in the face of maternal antiubody. Vaccine 38, 298-308.

Korhonen, H., Marnila, P., Gill, H.S. (2000). Milk immunoglobulins and complement factors. Br J Nutr 84 Suppl 1, S75-80.

Kühnle, G., Heinze, A., Schmitt, J., Giesow, K., Taylor, G., Morrison, I., Rijsewijk, F.A., van Oirschot, J.T., Keil, G.M. (1998). The class II membrane glycoprotein G of bovine respiratory syncytial virus, expressed from a synthetic open reading frame, is incorporated into virions of recombinant bovine herpesvirus 1. J Virol 72(5), 3804-11.

Kumar, P., Ayalew, L.E., Godson, D.L., Gaba, A., Babiuk, L.A., Tikoo, S.K. (2014). Mucosal immunization of calves with recombinant bovine adenovirus-3 expressing truncated BHV-1 gD and bovine IL-6. Vaccine 32(26), 3300-6.

Kunkel, E.J., Butcher, E. (2002). Chemokines and the Tissue-Specific Migration of Lymphocytes. Immunology 16, 1-4.

Larson, R.L., Step, D.L. (2012). Evidence-Based Effectiveness of Vaccination Against *Mannheimia haemolytica, Pasteurella multocida*, and *Histophilus somni* in Feedlot Cattle for Mitigating the Incidence and Effect of Bovine Respiratory Disease Complex. Vet Clin Food Anim 28, 97–106.

Lawan, A., Jessea, F.F.A., Idrisa, U.H., Odhah, M.N., Arsalanh, M., Muhammad, N.A., Bhuttoa, K.R., Peter, I.D., Abraham, G.A., Wahid, A.H., Mohd-Azmi, M.L., Zamri-Saadd, M. (2018). Mucosal and systemic responses of immunogenic vaccines candidates against enteric Escherichia coli infections in ruminants: A review. Microb Pathog 117, 175-183.

Lee, R.W., Cornelisse, M., Ziauddin, A., Slack, P.J., Hodgins, D.C., Strommer, J.N., Shewen, P.E., Lo, R.Y. (2008). Expression of a modified Mannheimia haemolytica GS60 outer membrane lipoprotein in transgenic alfalfa for the development of an edible vaccine against bovine pneumonic pasteurellosis. J Biotechnol 135(2), 224-31.

Lee, R.W., Strommer, J., Hodgins, D., Shewen, P.E., Niu, Y., Lo, R.Y. (2001). Towards development of an edible vaccine against bovine pneumonic pasteurellosis using transgenic white clover expressing a Mannheimia haemolytica A1 leukotoxin 50 fusion protein. Infect Immun 69(9), 5786-93.

Leigh, J.A. (1999). Streptococcus uberis: a permanent barrier to the control of bovine mastitis? Vet J 157(3), 225-38.

Lemaire, M., Meyer, G., Baranowski, E., Schynts, F., Wellemans, G., Kerkhofs, P., Thiry, E. (2000). Production of bovine herpesvirus type 1-seronegative latent carriers by administration of a live-attenuated vaccine in passively immunized calves. J Clin Microbiol 38(11), 4233-38.

Liang, G., Bao, H., McFadden, T., Stothard, P., Griebel, P., Guan, L.L. (2014). Potential regulatory role of microRNAs in the development of bovine gastrointestinal tract during early life. PLoS One 9(3), e92592.

Liang, X.P., Lamm, M.E., Nedrud, J.G. (1989). Cholera toxin as a mucosal adjuvant. Glutaraldehyde treatment dissociates adjuvanticity from toxicity. J Immunol 143(2), 484-90.

Liang, G., Malmuthuge, N., Guan, Y., Ren, Y., Griebel, P.J., Guan, L.L. (2016). Altered microRNA expression and pre-mRNA splicing events reveal new mechanisms associated with early stage Mycobacterium avium subspecies paratuberculosis infection. Sci Rep-Nature 6, 24964.

Liebler-Tenorio, E., Pabst, R., (2006). MALT structure and function in farm animals. Vet Res 37, 257-280.

Liljavirta, J., Ekman, A., Knight, J.S., Pernthaner, A., Iivanainen, A., Niku, M. (2013). Activation-induced cytidine deaminase (AID) is strongly expressed in the fetal bovine ileal Peyer's patch and spleen and is associated with expansion of the primary antibody repertoire in the absence of exogenous antigens. Mucosal Immunol 6(5), 942-9.

Lima, S.F., Teixeira, A.G., Higgins, C.H., Lima, F.S., Bicalho, R.C. (2016). The upper respiratory tract microbiome and its potential role in bovine respiratory disease and otitis media. Sci Rep 6, 29050.

Loehr, B.I., Rankin, R., Pontarollo, R., King, T., Willson, P., Babiuk, L.A., van Drunen Littel-van den Hurk, S. (2001). Suppository-mediated DNA immunization induces mucosal immunity against bovine herpesvirus-1 in cattle. Virology 289(2), 327-33.

Malmuthuge, N., Liang, G., Griebel, P.J., Guan, L.L. (2019). Taxonomic and Functional Compositions of the Small Intestinal Microbiome in Neonatal Calves Provide a Framework for Understanding Early Life Gut Health. Appl Environ Microbiol 85(6), e02534-18.

March, J.P., Prahud, J.J. (1971). Secretory IgA, a major immunoglobulin in most bovine external secretions. J Immunol 106, 552-563.

McCarville, J.L., Ayres, J.S. (2018). Disease Tolerance: Concept and Mechanisms. Curr Opin Immunol 50, 88-93.

McGhee, J.R., Mestecky, J. (1990). In Defence of Mucosal Surfaces. Development of Novel Vaccines for IgA Responses Protective at the Portals of Entry of Microbial Pathogens. Infect Dis Clin North Am 4(2), 319-41.

Mebus, C.A., White, R.G., Stair, E.L., Rhodes, M.B., Twiehaus, M.J. (1972). Neonatal calf diarrhea: results of a field trial using a reo-like virus vaccine. Vet Med Small Animal Clin 67, 173-74.

Midla, L.T., Hill, K.L., Van Engen, N.K., Edmonds, M., Renter, D.G., Streeter, M.N., Hutcheson, J.P., Griebel, P.J. (2020). Innate and acquired immune responses following intranasal vaccination with two commercially available modified-live viral vaccines in colostrum fed neonatal Holstein calves. JAVMA (In Press).

Miles, D.G. (2009). Overview of the North American beef cattle industry and the incidence of bovine respiratory disease (BRD). AHRR 10, 101-103.

Mirabzadeh-Ardakani, A., Solie, J., Gonzalez-Cano, P., Schmutz, S.M., Griebel, P. (2016). Expression of DEFB103 gene and protein. Cell Tissue Research 363, 479-490.

Newsted, D., Fallahi, F., Golshani, A., Azizi, A.(2015). Advances and Challenges in Mucosal Adjuvant Technology. Vaccine 33(21), 2399-405.

O'Connor, A., Cooper, V., Censi, L., Meyer, E., Kneipp, M., Dewell, G. (2019). A 2-year randomized blinded controlled trial of a conditionally licensed Moraxella bovoculi vaccine to aid in prevention of infectious bovine keratoconjunctivitis in Angus beef calves. J Vet Intern Med 33(6), 2786–2793.

O'Connor, A.M., Shen, H.G., Wang, C., Opriessnig, T. (2012). Descriptive epidemiology of Moraxella bovis, Moraxella bovoculi and Moraxella ovis in beef calves with naturally occurring infectious bovine keratoconjunctivitis (Pinkeye). Vet Microbiol 155(2-4), 374-380.

Okeke, E.B., Uzonna, J.E. (2019). The pivotal role of regulatory T cells in the regulation of innate immune cells. Front Immunol 10, 680.

Osman, R., Gonzalez-Cano, P., Brownlie, R., Griebel, P. (2017). Induction of Interferon and Interferon-Induced Antiviral Effector Genes Following a Primary Bovine Herpes Virus-1 (BHV-1) Respiratory Infection. J. Gen Virol 98(7), 1831-1842.

Osman, R., Malmuthuge, N., Gonzalez-Cano, P., Griebel, P. (2018). Development and Function of the Mucosal Immune System in the Upper Respiratory Tract of Neonatal Calves. Ann Rev Animal Biosciences 6,141-155.

Pavot, V., Rochereau, N., Genin, C., Verrier, B., Paul, S. (2012). New insights in mucosal vaccine development. Vaccine 30, 142-154.

Peterson, L.W., Artis, D. (2014). Intestinal epithelial cells: regulators of barrier function and immune homeostasis. Nature Reviews Immunology 14, 141-153.

Prysliak, T., van der Merwe, J., Lawman, Z., Wilson, D., Townsend, H., van Drunen Littel-van den Hurk, S., Perez-Casal, J. (2011). Respiratory disease caused by Mycoplasma bovis is enhanced by exposure to bovine herpes virus 1 (BHV-1) but not to bovine viral diarrhea virus (BVDV) type 2. CVJ 52, 1195-202.

Pugh, G.W. Jr, Kopecky, K.E., McDonald, T.J. Infectious bovine keratoconjunctivitis: subconjunctival administration of a Moraxella bovis pilus preparation enhances immunogenicity. Am J Vet Res 46(4), 811-15.

Rainard, P., Foucras, G., Fitzgerald, J.R., Watts, J.L., Koop, G., Middleton, J.R. (2018). Knowledge gaps and research priorities in Staphylococcus aureus mastitis control. Transbound Emerg Dis 65 Suppl 1, 149-165.

Ribble, C.S., Meek, A.H., Shewen, P.E., Jim, G.K., Guichon, P.T. (1995). Effect of transportation on fatal fibrinous pneumonia and shrinkage in calves arriving at a large feedlot. JAVMA 207, 612-615.

Ridpath, J.E., Neill, J.D., Endsley, J., Roth, J.A. (2003). Effect of passive immunity on the development of a protective immune response against bovine viral diarrhea virus in calves. Am J Vet Res 64, 65-69.

Rosenquist, B.D., Loan, R.W. (19676). Interferon production with strain SF-4 of parainfluenza 3 virus. Am J Vet Res 32, 35-39.

Roussey, J.A., Oliveira, L.J., Langohr, I.M., Sledge, D.G., Coussens, P.M. (2016). Regulatory T cells and immune profiling in Johne's disease lesions. Vet Immunol Immunopathol 181, 39-50. Roussey, J.A., Steibel, J.P., Coussens, P.M. (2014). Regulatory T Cell Activity and Signs of T Cell Unresponsiveness in Bovine Paratuberculosis. Front Vet Sci 1: 20.

Sanchis, R., Abadie, G., Pardon, P. (1995). Salmonella abortusovis experimental infection induced by the conjunctival route: clinical, serological and bacteriological study of the dose effect in female lambs. Vet. Res 26, 73-80.

Savan, M., Angulo, A.B., Derbyshire, J.B. (1979). Interferon, Antibody Responses and Protection Induced by an Intranasal Infectious Bovine Rhinotracheitis Vaccine. Can Vet J 20, 207-210.

Slocombe R.F., Malark, J., Ingersoll, R., Derksen, F.J., Robinson, N.E. (1985). Importance of neutrophils in the pathogenesis of acute pneumonic pasteurellosis in calves. Am J Vet Res 46, 2253-2258.

Smith, M.W., Miller,R.B., Svoboda, I., Lawson K.F. (1978). Efficacy of an intranasal infectious bovine rhinotracheitis vaccine for the prevention of abortion in cattle. Can Vet J 19(3), 63–71.

Sokolowska, M., Frei, R., Lunjani, N., Akdis, C.A., O'Mahony, L. (2018). Microbiome and asthma. Asthma Res Pract 4, 1.

Sordillo, L.M. (2018). Mammary Gland Immunobiology and Resistance to Mastitis. Vet Clin North Am Food Anim Pract 34(3), 507-523.

Step, D.L., Krehbiel, C.R., DePra, H.A., Cranston, J.J., Fulton, R.W., Kirkpatrick, J.G. (2008). Effects of commingling beef calves from different sources and weaning protocols during a forty-two-day receiving period on performance and bovine respiratory disease. J Anim Sci 86, 3146-3158.

Taschuk, R., Griebel, P.J. (2012). Commensal microbiome effects on mucosal immune system development in the ruminant gastrointestinal tract. AHRR 13(1), 129-141.

Tewari, D., Hovingh, E., Linscott, R., Martel, E., Lawrence, J., Wolfgang, D., Griswold, D. (2014). Mycobacterium *avium* subsp. *paratuberculosis* antibody response, fecal shedding, and antibody cross-reactivity to Mycobacterium bovis in *M. avium* subsp. *paratuberculosis*-infected cattle herds vaccinated against Johne's disease. Clin Vaccine Immunol 21(5), 698-703.

Todd, J.D. (1974). Development of intranasal vaccination for the immunization of cattle against infectious bovine rhinotracheitis. Can Vet J 15(9), 257–259.

Todd, J.D., Volenec, F.J., Paton, I.M. (1971). Intranasal vaccination against infectious bovine rhinotracheitis: studies on early onset of protection and use of the vaccine in pregnant cows. JAVMA 159(11), 1370-1374.

Todd, J,D., Volenec, F.J., Paton, I.M. (1972). Interferon in Nasal Secretions and Sera of Calves After Intranasal Administration of Avirulent Infectious Bovine Rhinotracheitis Virus: Association of Interferon in Nasal Secretions with Early Resistance to Challenge with Virulent Virus. Infect Immun 5(5), 699–706.

Van Zaane, D., Ijzerman, J., De Leeuw, P.W. (1986). Intestinal antibody response after vaccination and infection with rotavirus of calves fed colostrum with or without rotavirus antibody. Vet Immunol Immunopathol 11(1), 45–63.

Walz, P.H., Newcomer, B.W., Riddell, K.P., Scruggs, D.W., Cortese, V.S. (2017). Virus detection by PCR following vaccination of naive calves with intranasal or injectable multivalent modified-live viral vaccines. J Vet Diagn Invest 29(5), 628-635.

Weikard, R., Demasius, W., Hadlich, F., Kühn, C. (2015). Different Blood Cell-Derived Transcriptome Signatures in Cows Exposed to Vaccination Pre- or Postpartum. PLoS One 10(8), e0136927.

Wiedle, G., Dunon, D., Imhof, B.A. (2001). Current Concepts in Lymphocyte Homing and Recirculation, Critical Reviews in Clinical Laboratory Sciences 38, 1-31.

Williams, L.B.A., Fry, L.M., Herndon, D.R., Franceschi, V., Schneider, D.A., Donofrio, G., Knowles, D.P. (2019). A recombinant bovine herpesvirus-4 vectored vaccine delivered via intranasal nebulization elicits viral neutralizing antibody titers in cattle. PLoS One 14(4), e0215605.

Woolums, A.R., Siger, L., Johnson, S., Gallo, G., Conlon, J. (2003). Rapid onset of protection following vaccination of calves with multivalent vaccines containing modified-live or modified-live and killed BHV-1 is associated with virus-specific interferon gamma production. Vaccine 21:1158-64.

Xue, W., Ellis, J., Mattick, D., Smith, L., Brady, R., Trigo, E. (2010). Immunogenicity of a modified-live virus vaccine against bovine viral diarrhea virus types 1 and 2, infectious bovine rhinotracheitis virus, bovine parainfluenza-3 virus, and bovine respiratory syncytial virus when administered intranasally in young calves. Vaccine 28(22), 3784-92.

Xue, W., Mattick, D., Smith, L. (2011). Protection from persistent infection with a bovine viral diarrhea virus (BVDV) type 1b strain by a modified-live vaccine containing BVDV types 1a and 2, infectious bovine rhinotracheitis virus, parainfluenza 3 virus and bovine respiratory syncytial virus. Vaccine 29(29-30), 4657-62.

Yates, W.D. (1982). A review of infectious bovine rhinotracheitis, shipping fever pneumonia and viral-bacterial synergism in respiratory disease of cattle. Can J Comp Med 46(3), 225-63.

Yates, W.D.G., Babiuk, L.A., Jericho, K.W.F. (1983). Viral–bacterial pneumonia in calves: duration of the interaction between bovine herpesvirus 1 and Pasteurella haemolytica. Can J Comp Med 47, 257-26.

Zbrun, M.V., Zielinski, G.C., Piscitelli, H.C., Descarga, C., Urbani, L.A., Defain, C., Tesoriero, M.V., Hermida, L. (2012). Evaluation of anti-Moraxella bovis pili immunoglobulin-A in tears following intranasal vaccination of cattle. Res Vet Sci 93(1), 183-9.

Chapter 4

Adjuvants: State of the Art and New Developments

Sylvia van Drunen Littel-van den Hurk[1,2,*] and George Mutwiri[2,3]

[1]VIDO-InterVac, University of Saskatchewan, Saskatoon, SK, S7N 5E3, Canada
[2]Microbiology and Immunology, University of Saskatchewan, Saskatoon, SK, S7N 5E3, Canada
[3]School of Public Health, University of Saskatchewan, Saskatoon, SK, S7N 5E3, Canada

*Corresponding author: sylvia.vandenhurk@usask.ca

DOI: https://doi.org/10.21775/9781913652593.04

Abstract

Adjuvants are used for formulation of vaccine antigens to increase immune responses. While only a few different kinds of adjuvant are used in veterinary vaccines, many new adjuvant types are under investigation, and although much research still needs to be done, we have increased knowledge of the mechanism of action of adjuvants. We also have a much better understanding of the innate immune system, and the link with adaptive immunity. This allows us to develop a better rationale for selection of adjuvants suitable for specific antigens, diseases, animal species and routes of delivery. Since multiple receptors and ligands are involved in activation of the innate immune response, adjuvant formulations can be generated that contain one or several of those ligands in combinations that are tailor-made for induction of humoral, cell-mediated or balanced immune responses. Many new adjuvants combine several mechanisms of action, which can be additive or synergistic. In this chapter we discuss the state of the art and recent progress in adjuvant development, particularly in the context of veterinary vaccines, as well as the mechanism of action known for some of the adjuvants.

Introduction

Adjuvants are substances that increase immune responses to antigens (Ramon, 1924), and have been used to enhance immunogenicity and efficacy of vaccines for decades. While the original definition may have been limited to increasing antibody titers, adjuvants now have a broader definition, which includes modulation of the quantity and quality of immune responses. In this regard, adjuvants can modulate antibody avidity, specificity, isotype and subclass distribution (Rajput et al., 2007) and modulate Th1, Th2 or Th17 types of immune responses (Awate et al., 2013). Consequently, adjuvants can have various functions, namely: i) reduce the amount of antigen per dose; ii) reduce the number of immunizations required for protection; iii) promote early onset of immunity; iv) extend duration of protective immunity; v) improve mucosal vaccines; vi) tailor vaccines to induce Th1-, Th2- and/or Th17- type responses, and vii) overall, improve vaccines for newborns, elderly and immunocompromised (Rajput et al., 2007).

Many of the veterinary vaccines are modified-live organisms, which mimic natural infection and therefore stimulate appropriate, often strong, immune responses and usually do not require adjuvants. However, modified-live vaccines (MLV) have safety concerns, especially in neonates and immunocompromised, including pregnant animals. While safer than MLVs, killed vaccines are less immunogenic, although these vaccines may contain components that possess immunostimulatory properties and/or be formulated with adjuvants. In contrast, subunit vaccines consisting of purified antigens generally are poorly immunogenic unless they are formulated with adjuvants. With the ability to identify protective antigens, by reverse vaccinology if necessary, and to generate recombinant proteins, subunit vaccines are more and more attractive as safe, defined, possibly DIVA (differentiation of vaccinated from infected animals), vaccines,and thus the selection and formulation with potent, appropriate adjuvants are becoming more important, both for human and veterinary vaccines. Adjuvants are also being tested as an approach to enhance the efficacy of vectored vaccines.

Only a limited number of adjuvants are used in licensed veterinary vaccines, including Freund's incomplete adjuvant, Montanide, aluminum hydroxide, potassium aluminum sulfate (alum), aluminum phosphate (Alhydrogel), Emulsigen® D (MVP Technologies), QS-21 and Matrix-M (Chambers et al., 2016). In contrast to the limited number of adjuvants that are approved as part of licensed vaccines, a plethora of adjuvants of various compositions and with diverse actions is currently in pre-clinical and clinical development. The state of the art and recent progress in the context of veterinary vaccines will be discussed in this chapter, with respect to the adjuvant composition, activity and mechanism of action where known.

State of the art and recent developments in adjuvants for veterinary use

A variety of adjuvant types with different mechanisms of action is currently in development for use in veterinary vaccines. Based on their composition and action they can be categorized as follows: i) mineral salts, ii) oil emulsions (oil-in water/water-in-oil), iii) toll-like receptor (TLR)/nucleotide-binding oligomerization domain (NOD)-like receptor (NLR) agonists, iv) cytokines/chemokines, v) synthetic polymers, vi) saponins and vii) delivery systems with or without adjuvant activity, i.e. ISCOMs, micro/nanoparticles, and liposomes/virosomes. These adjuvants have unique and sometimes overlapping mechanisms of action. Several of the most promising adjuvants are combinations of compounds with complementary or synergizing properties (Levast et al., 2014).

Mineral Salts

Aluminum salt is an adjuvant commonly used in human and veterinary vaccines Several types with slightly different properties exist, namely potassium aluminum sulfate (Alum), aluminum hydroxide (Alhydrogel), aluminum phosphate (Adju-Plus) and aluminum hydroxide-magnesium hydroxide (Imject Alum). These adjuvants have been used for decades in many human and animal vaccines and are generally very safe. One exception has been a correlation between vaccine-associated sarcomas observed at the site of vaccination in cats and the use of aluminum adjuvants, because residual aluminum was observed in the sarcomas (Kass et al., 1993). Aluminum-based adjuvants induce Th2-biased, humoral immune responses, which are protective against most extracellular pathogens, but are ineffective against many viruses and other intracellular organisms. Therefore, Alum has been recently evaluated in combination adjuvant formulations, for example with a Th1-stimulating compound such as CpG oligodeoxynucleotides (ODNs). Such combinations, discussed in more detail in subsequent sections, may elicit a balanced immune response, for example to bovine herpesvirus-1 (BoHV-1) glycoprotein D (gD) or *Anaplasma marginale* major surface protein 2 in cattle (Rankin et al., 2002; Zhang et al., 2003). The mechanism of action (MOA) of aluminum salts is not entirely understood, but may involve depot/slow release, enhancement of antigen uptake by antigen presenting cells (APCs) and direct or indirect activation of the NACHT, LRR and PYD domains-containing protein 3 (NLRP3) inflammasome by the aluminum salt itself and/or the release of damage-associated molecular patterns (DAMPs) due to tissue injury (Oleszycka and Lavelle, 2014). In bovine peripheral blood mononuclear cells (PBMCs) Alum was shown to enhance IL-1β secretion in a NLRP3-, caspase-1- and caspase-8 – dependent manner. It was proposed that after cell entry Alum induces lysosomal degradation which releases cathepsin; this may drive NLRP3 activation, ASC oligmerization and caspase-1 and/or caspase -8 activation, and ultimately release of IL-1β and IL-18 (Harte et al., 2017). For detailed information about Alum and its MOA two excellent reviews can be consulted (HogenEsch et al., 2018; Kool et al., 2012).

Oil emulsions

Oil-in-water and water-in oil are emulsion-type adjuvants, consisting of micro-size uniform oil or water droplets; the presence and type of surfactants in the formulation control the formation and size of the droplets. While the well-known Freund's adjuvants (complete and incomplete) are very effective and still used in laboratory settings for production of antibodies, a major disadvantage is the induction of inflammation and granulomas (Stills, 2005). Several oil emulsion adjuvants are available for use in vaccines for livestock species, including MF59 (Novartis), Emulsigen® D (MVP Technologies), Syntex adjuvant formulation (SAF, SAF-m, SAF-i), and Montanide – ISA W/O (Seppic). These adjuvants form an emulsion with the aqueous phase containing the antigen, and mainly function by causing a depot effect and attracting and activating APCs (Singh and O'Hagan, 2003).

A number of new oil-emulsion adjuvants was recently evaluated in pigs. While a commercial killed porcine respiratory and reproductive syndrome virus (PRRSV) vaccine failed to elicit protection, binary ethyleneimine (BEI)-inactivated PRRSV vaccine formulated with Montanide Gel 01 ST generated cross-protective virus neutralizing (VN) antibodies to both homologous and heterologous PRRSV strains in pigs (Tabynov et al., 2016). Several adjuvants were compared in a formulation with a BEI-inactivated foot and mouth virus (FMDV) vaccine: new adjuvants Emulsigen® D, Carbigen (MVP Technologies), and Montanide ISA 201 and well-known adjuvants

ISA 206 and Alhydrogel. This study showed Emulsigen® D and ISA 201 to be superior adjuvants mediating stronger VN responses and better protection of pigs from FMDV challenge (Park et al., 2014). A new oil-in-water emulsion, OW-14, containing low-cost plant-based emulsifiers was tested for its efficacy with whole killed swine influenza virus (SIV) and whole killed *Mycoplasma hyopneumoniae* bacterin in comparison to the commercial FluSure (SIV) or RespiSure (*M. hyo*) vaccines. OW-14-adjuvanted inactivated SIVs (H3N2 and H1N1) and *M. hyo* vaccines were safe and effective. Vaccinated pigs developed high and prolonged antibody titers to both SIV and *M. hyo*, either comparable or greater than those produced by FluSure or RespiSure vaccines. This adjuvant was inexpensive and easy to use as it could be simply mixed with the antigens (Galliher-Beckley et al., 2015).

Several Montanide adjuvants with a variety of antigens have been tested in ruminants. While formulation of inactivated FMDV with Montanide™ IMS 1313 N VG NPR resulted in protection in 60% of vaccinated calves, formulation with Montanide™ ESSAI IMS D 12802 VG PR (802-iFMDV) elicited 100% protection in intramuscularly (IM)-vaccinated cattle. Interestingly, the numbers of IFN-γ secreting PBMCs and nasal IgA and IgG1 were all increased, but the VN antibody levels were low. In contrast, the opsonophagocytosis of virus was significantly increased in the 802-iFMDV group. This suggests that the increase in the activated monocytes numbers and the presence of antibodies with opsonizing activity in these animals mediated protection (Quattrocchi et al., 2014). Montanide™ ISA 206 VG has been tested in formulations with the paramyosin rSj97 of *Schistosoma japonicum* and the sporozoite protein p67 of *Theileria parva*. Three IM doses of adjuvanted rSj97 elicited both IgG1 and IgG2 production and partial protection from *S. japonicum* in water buffalo (Wu et al., 2017). Partial protection of cattle from *T. parva* was achieved with three subcutaneous (SC) p67 vaccine doses, and preliminary correlates of protection with CD4 T-cell proliferation index and IgG, IgG1 and IgG2 titers were identified (Lacasta et al., 2018). Interestingly, the Montanide™ adjuvants Essai 903110 and 1113101 PR were evaluated with an intradermally (ID) administered BoHV-1 gD DNA vaccine (three doses). Both adjuvants increased humoral immune responses and reduced clinical symptoms, while only Montanide™ 1113101 PR enhanced cell-mediated immune (CMI) responses and reduced virus replication, although significant virus shedding was still apparent in the vaccinated animals (Quattrocchi et al., 2017).

Toll-like receptor and NOD-like receptor agonists.

The innate immune responses to pathogens are initiated by recognition of pathogen-associated molecular patters (PAMPs) by pattern recognition receptors (PRRs), including TLRs and NLRs. Modified-live vaccines usually activate multiple TLRs; for instance, the highly effective yellow fever vaccine activates TLR2, 7, 8 and 9 (Querec et al., 2006). The heat-killed mycobacteria in Freund's complete adjuvant also act through multiple TLRs. Based on this information, synthetic compounds that recognize and activate TLRs and NLRs are now in use as vaccine adjuvants. Activation of the PRRs by their agonists leads to expression of pro-inflammatory cytokines and activation of APCs through multiple signaling pathways. Depending on the PAMP, Th1-type, Th2-type or balanced immune responses are induced as described in detail in a review by Toussi and Masari (Toussi and Massari, 2014), which renders these compounds very suitable for tailor-made adjuvanting. For an excellent review on NLR and TLR agonists in context of instruction of adaptive immune responses and tailoring to specific pathogens the reader is referred to Maisonneuve et al. (Maisonneuve et al., 2014).

Synthetic nucleic acids are PAMPs that induce type I IFN and IL-12 production, and Th1-biased responses, including cytotoxic T cells (CTLs). For example, poly(I:C) (dsRNA) is a ligand for endosomal TLR3, cytosolic retinoic acid-inducible gene-I (RIG-I) and melanoma-differentiation associated protein 5 (MDA5). Its derivatives, polyI:C$_{12}$U (Ampligen) and poly(IC:LC) (Hiltonol), are used in clinical trials against HIV (Toussi and Massari, 2014). TLR7 and 8 recognize ssRNA, as well as small-molecule potentiators (SMIPs) such as Imiquimod and Resiquimod, which are used in human papilloma virus (HPV) vaccines (Toussi and Massari, 2014). TLR9 recognizes unmethylated CpG motifs in dsDNA. The TLR9 agonist CpG ODN has been most often tested as adjuvant, with multiple antigens, and is in use with a commercial hepatitis B virus (HBV) vaccine (Toussi and Massari, 2014). TLR9 signaling leads to Th1-type pro-inflammatory responses including IL-1, IL-6, IL-12, TNF-α and IFN-γ, upregulation of MHC, followed by increased CD8 T cell responses and antibody class switching and humoral responses. Synthetic dsDNAs can also be recognized by other PRRs such as absent in melanoma2 (AIM2), as well as by the protein cyclic guanosine monophosphate-adenosine monophosphate (cGAMP) synthase (cGAS), which leads to the activation of the stimulator of IFN genes (STING)-dependent pathway. The natural agonist of STING, cGAMP, has been used as adjuvant (Christensen and Paludan, 2017).

Bacterial PAMPs recognize cell surface PRRs, and elicit either Th1 or Th2-biased immune responses. Lipopolysaccharide (LPS) and its much less toxic derivative monophosphoryl lipid A (MPL) bind to TLR4 and promote a Th-1 biased response; MPL is one of the components in AS04, the adjuvant in HBV and HPV vaccines, as well as in AS01, the adjuvant in VZV and malaria vaccines (Toussi and Massari, 2014). Recognition and activation of TLR5 by bacterial flagellin leads to Th1, Th2 or mixed responses in mice (Hajam et al., 2017). The adjuvanticity of TLR2 is mostly Th2-biased. TLR2-TLR1 complexes are activated by lipopeptide analog Pam_3CSK_4, while TLR2-TLR6 complexes recognize macrophage activating lipopolypeptide-1 (MAPLP-2) from mycoplasma (Toussi and Massari, 2014).

When used as adjuvants, synthetic PAMPs are often formulated with other compounds as combination adjuvant, in order to protect them from early degradation or elimination, and facilitate cell uptake. For example, cell entry of poly(I:C) or CpG ODN that interact with endosomal (TLR3,7,8 and 9) or cytosolic (RIG-I and MDA5) PRRs can be enhanced by formulation with cationic phospholipids. Several well-known adjuvants such as Alum, MF59, Cholera Toxin, Quil A and chitosan either act as, or induce the production of DAMPs. The DAMPs are recognized by NLRs, such as NLRP3. This results in activation of the inflammasome and caspases, followed by production of pro-inflammatory cytokines IL-1β and IL-18 (Aoshi, 2017).

Many TLR agonists are under investigation as adjuvants in veterinary vaccines. Recently, TLR agonists have been used to adjuvant a replication - defective adenovirus 5 vector carrying the FMDV capsid. The addition of polyICLC resulted in a 80-fold reduction of the protective dose, while still achieving protection in the absence of VN antibodies in pigs (Diaz-San Segundo et al., 2014). In another interesting approach synthetic non-infectious RNA derived from the non-coding regions of the FMDV genome (ncRNAs) was used to adjuvant a conventional FMDV vaccine, which resulted in earlier and higher VN antibody levels specific for homologous and heterologous isolates, higher antigen-specific T cells responses and increased protection levels (Borrego et al., 2017). Recently, an interesting role of TLR8 in detection of live vs. dead bacteria was identified in pigs and humans. Human APCs distinguished viable bacteria from dead bacteria through TLR8-dependent detection of bacterial RNA. The recognition of bacterial viability by the innate immune system led to transcriptional remodeling in human APCs and induced Tfollicular helper (T_{FH} cell) - promoting signals, most importantly IL-12. In pigs, immunization with a live Salmonella vaccine, Salmoporc-STM, induced robust T_{FH} cell and antibody responses, whereas immunization with its heat-killed counterpart did not. TLR8 agonists may thus function in promotion of T_{FH} cell skewing responses (Ugolini et al., 2018).

A variety of TLR agonists have been tested in chickens. Flagellin has recently been of particular interest (reviewed in (Gupta et al., 2014)). In chickens it elicits a balanced immune response, with a slight bias towards Th2. Flagellin is effective at very low doses, and being a protein, can be linked to the antigen of interest. Flagellin as adjuvant has been evaluated in chickens with antigens of *Eimeria tenella*, *Salmonella enteritica, Campylobacter jejuni* and avian influenza virus, resulting in increased IgA and/or IgY, and reduced bacterial counts when challenged (Gupta et al., 2014). The TLR8 agonist, Resiquimod, also has adjuvant activity in chickens. When formulated with inactivated Newcastle disease virus (NDV), Resiquimod up-regulated the expression of several immune response genes including IFN-α, IFN-β, IFN-γ, IL-4, IL-1β, MHC-II and iNOS in chicken spleens. Furthermore, antibody levels and CD4 and CD8 T cells were increased, and all chickens were completely protected from NDV challenge (Sachan et al., 2015). When Resiquimod was formulated with killed or live infectious bronchitis virus (IBV) vaccine and administered IM to chickens, both IgG and IgA, as well as CD4 and CD8 T cells, were upregulated, while the major cytokine was TGF-β4 known to mediate class switching of B cells to produce IgA (Matoo et al., 2018). CpG ODN has been frequently tested as adjuvant in chickens. In a recent trial chicken infectious anemia virus (CIAV) VP1, VP2 and VP3 proteins induced higher CIAV-specific antibody titers when formulated with CpG ODN than with Freund's adjuvant (Fang et al., 2019).

Cytokines

Since cytokines can modulate immune responses, they have been evaluated as adjuvants. Many cytokines, most frequently IL-1, IL-12, IFN-γ and GM-CSF, have been tested, mostly in mouse models. IL-12 for example can be used to drive a Th1-biased immune response. While cytokines may have adjuvant potential, they have several disadvantages, including the high cost which is an impediment for veterinary use, the potential for side effects, and the short half-life. The use of TLR or NLR agonists as described above will also lead to cytokine induction without these disadvantages. However, incorporation of cytokine-encoding genes, either into plasmid DNA or into parasites, viruses or bacteria, would reduce the cost and enhance the availability to the immune system. In the context of veterinary vaccines, most recent research on cytokines has been performed in chickens.

For example, a transgenic *Eimeria mitis* expressing chicken IL-2 (EmiChIL-2) was constructed; IL-2 is a canonical T cell growth factor and plays a role in the expansion and differentiation of CD4 T cells. Compared to wild-type Eimeria, EmiChIL-2 elicited similar humoral and stronger CMI responses. When the immunized chickens were challenged with the parasite, EmiChIL-2 immunized chickens excreted fewer oocysts as compared with wild type-immunized chickens correlating to the enhanced immune responses induced by ChIL-2 (Li et al., 2015). To test the ability of cytokines to assist in eliciting antibody responses in chicks with maternal antibodies, ChIL-2, ChIL-15, ChGM-CSF, or fliC (flagellin), were expressed in NDV. Specific pathogen-free chickens immunized with any one of these recombinant viruses developed more rapid and significantly higher haemagglutination inhibition (HI) and VN antibody titers compared to the parental virus. Furthermore, NDV expressing ChIL-2, ChIL-15, ChGM-CSF or fliC promoted CD4 T and CD8 T proliferation. The recombinant viruses elicited 100% protection, while the parental virus induced 80% protection. Importantly, in the presence of maternal antibody the parental virus did not elicit protective HI titers until day 14, but NDV expressing GM-CSF or fliC elicited protective antibody levels by day 7 post vaccination (Zhang et al., 2016b). A delayed lysis Salmonella strain was used to express the fusion protein from NDV either alone or together with chicken IL-18 (ChIL-18). Like the mammalian IL-18, ChIL−18 induces production of IFN−γ. Oral immunization of either recombinant strain stimulated enhanced humoral and cellular immune responses in chickens and provided partial protection against wild−type intranasal NDV challenge; however, the lymphocyte proliferative response was higher in the chickens that received IL-18 in addition to the fusion protein, and the number of protected chickens was higher suggesting that this may be a promising vaccine candidate against genotype VII NDV (Gao et al., 2019).

Synthetic polymers

A variety of synthetic polymers, including dextrans, polyesther amines, polyesther, polyethylenimine, polyanhydrides, polysaccharides, polyacrylic acid and polyphosphazenes, can be used as adjuvant. Carbopol-based adjuvants, used in veterinary vaccines since the 1970's, have been shown to be safe and effective in all animal species tested. Carbopol® homopolymers are acrylic acid crosslinked with allyl sucrose or allyl pentaerythritol, and Carbopol® copolymers consist of acrylic acid and C10-C30 alkyl acrylate crosslinked with allyl pentaerythritol. As a recent example, Carbopol, when formulated with a PRRS MLV (Ingelvac), was shown to enhance the number of early IFN-γ secreting cells and to drive T cell differentiation into effector phenotypes in piglets (Mair et al., 2015). One of the carbomers, Carbopol-941P, is a component in a licensed vaccine for pigs (the classical swine fever (CSF) DIVA vaccine; Suvaxyn, Pfizer) (Blome et al., 2017). In cattle, *Moraxella bovis* cytotoxin (MbxA) formulated with Carbigen (MVP Technologies) was administerd intranasally. Changes in tear antigen-specific IgA levels from day 0 to days 28, 42, and 55 were significantly different between vaccinated and control groups (Angelos et al., 2014).

Polyethyleneamine (PEI) is extensively used as *in vitro* transfection agent, and is also effective as adjuvant in chickens. When formulated with influenza HA, PEI mediated increased IgG titers, percentages of CD4 and CD8 T cells, and upregulation of IFN-γ and IL-4 production by splenocytes (Song et al., 2018). The polysaccharides used as adjuvants mainly include chitosan and its derivatives, including glucan, mannan, inulin, and Chinese medicinal herb polysaccharides. Chitosan has been used in many applications; it displays low toxicity, environmental compatibility, biocompatibility, biodegradability, and muco-adhesive capacity. While chitosan is poorly water-soluble, chitosan quaternary ammonium salt is a water-soluble cationic polymer that retains the original properties of chitosan and can open tight intracellular junctions (Sun et al., 2018). Therefore, quaternized chitosan micro- or nanoparticles have been extensively tested in vaccine formulations. For example, in chickens intranasal administration of NDV-IBV MLV formulated with quaternized chitosan resulted in higher IgG and IgA levels and IFN-γ and IL-4 production when compared to MLV alone. The responses were protective against both NDV and IBV (Zhao et al., 2017). Many Chinese herb polysaccharides have been tested as adjuvants, mainly in chickens; several of them have been reviewed by Sun et al. (Sun et al., 2018). As an example not mentioned in this review, Taishan Pinus massonia pollen polysaccharide (TPPPS) was used to formulate a *Bordetella avium* ompA-DNA vaccine, and administered IM to chickens. The TPPPS-adjuvanted ompA DNA vaccine and Freund's adjuvanted ompA DNA vaccine formulations elicited stronger immune responses than the ompA DNA vaccine in PBS; however, only the ompA DNA vaccine with 200 mg/mL TPPPS completely protected the chickens against *B. avium* infection (Zhu et al., 2016).

Saponins

Saponins are natural glycosides of steroid or triterpene and found in plants (Sun et al., 2009). The most extensively studied saponin is derived from *Quillaja saponaria* Molina. An enriched extract containing a mixture of saponins termed Quil A was used in a FMDV vaccine (Dalsgaard, 1978). Quil A is also used in vaccines for cats, dogs, pigs, cattle and horses (Spickler and Roth, 2003). A major advantage of saponins is their ability to promote induction of both humoral and CMI responses. Since Quil A is quite toxic, individual fractions were purified showing that all of them have adjuvant activity, but QS-21 was found to be much less toxic than the other fractions (Kensil, 1996). QS-21 is the adjuvant in the commercially available feline leukemia virus vaccine (FeLV) (Adair, 2009), promoting production of Th1 cytokines, IgG2a and cytotoxic T cells. Saponin-based adjuvants stimulate influx of APCs into the injection site and enhance antigen presentation in the local draining lymph nodes (LNs).

Immune stimulating complexes (ISCOMs)

Immune stimulating complexes (ISCOMs), ISCOM-matrix and MATRIX-M are cage-like structures formed from saponin, cholesterol and phospholipid. In addition to having strong immunostimulatory properties that promote both humoral and CMI responses, these adjuvants act as delivery systems and are less toxic than Quil A. ISCOMS have been used with experimental vaccines against BoHV-1, bovine viral diarrhea virus (BVDV), rinderpest virus (RPV), canine distemper virus (CDV), pseudorabies virus (PRV) and FeLV in many animal species including cattle, dogs, pigs and cats (Sanders et al., 2005; Spickler and Roth, 2003). While the antigen is incorporated during the formation of ISCOMS, ISCOM-matrix is preformed before formulation with antigen.

Recently, ISCOM-matrix was formulated with a live attenuated *Mycoplasma hyopneumoniae* vaccine. Enhanced antigen-specific serum IgG and lymphocyte proliferation were observed, as well as reduced lesion scores after challenge of pigs with *M. hyopneumoniae* (Xiong et al., 2014). Matrix-M is a mixture of Matrix particles made from two different purified fractions of Quillaja-derived saponin. The local and systemic responses induced by Matrix-M were evaluated in piglets, suggesting activation of type I IFN through several signaling pathways (Ahlberg et al., 2017). After IM injection, Matrix-M induced leukocyte infiltration and upregulation of gene expression indicative of myeloid and plasmacytoid DCs. In addition, upregulation of cytokine/chemokine genes, mostly IFN-regulated genes (IRGs), as well as increased osteoporin, stimulator of IFN genes (STING) and INF-α and IFN-β production in the LNs was observed. This suggests that both the endosomal TLR7/TLR9 pathway including MyD88 and the cytosolic STING/TBK1 pathways are involved. Up-regulation of IFN-α and TLR2 was found in response to IM injection of Matrix-M in the blood (Fossum et al., 2014).

ISCOM Matrix is efficacious in cattle with several antigens. Recently, formulation of ISCOM Matrix with a *Staphylococcua aureus* CP5 whole lysate, alone or supplemented with defined *S. aureus* antigens, lead to induction of regulatory and pro-inflammatory cytokine production at the innate level, and strong antibody responses in blood and milk, which increased opsonic capacity of neutrophils, in vaccinated pregnant heifers (Camussone et al., 2014a; Camussone et al., 2014b).
The ISCOM, AbISCO®-300, was formulated with the 3-1E protein of *Eimeria acervulina* and administered IM to chickens. Higher CD3, CD4 and CD8 intraepithelial T cells and more intestinal IgA, as well as increased weight gain and decreased oocyst shedding, were observed in chickens immunized with the 3-1E protein formulated with AbISCO®-300 than in those vaccinated with the 3-1E protein alone (Zhang et al., 2014).

Micro- and nanoparticles and virus-like particles

Microparticles (2-5 µm) and nanoparticles (50-100 nm) (Shah et al., 2014) are composed of biodegradable and biocompatible materials, that can be either synthetic (polylactide-co-glycolides [PLGAs], poloxamers and polyphosphazenes [PPs]) or natural (chitosan, starch and alginates) (Gerdts et al., 2013). Micro- and nanoparticles promote uptake of antigens by APCs. Several types of particles have immunostimulatory properties, and induce both humoral and CMI responses including CTLs (Gerdts et al., 2013; Gregory et al., 2013). In addition to the antigens, immunomodulatory compounds such as CpG ODN, can be incorporated. An additional advantage is the protection from harsh environments in the respiratory and enteric tracts offered by micro/nanoparticles, making them suitable for mucosal (intranasal and oral) delivery (Gerdts et al., 2013; Singh and O'Hagan, 2003). PLGA has been used for many years as suture material and controlled release delivery systems, and is considered safe. The PLGA microparticles have been extensively studied for vaccine delivery, and can form a depot from which the antigen may be released slowly or in two bursts. However, the fact that they need to be prepared in organic solvents limits their use to antigens not sensitive to this treatment. In the context of veterinary vaccines, micro/nanoparticles are being developed and tested in cattle, swine, horses and fish.

Virus-like particles (VLPs) consist of one or several recombinant viral proteins that self-assemble into icosahedral or rod-like structures of 20-100 nm. VLPs act as vaccine antigen themselves and can function as a carrier of other antigens that can be genetically fused, incorporated or covalently linked. VLPs elicit strong antibody and moderate CMI responses, which may be further enhanced by incorporating TLR ligands such as ssRNA or CpG ODN. VLP-based vaccines are in development for bovine rotavirus, bluetongue virus, NDV, and avian influenza virus. Particles as delivery systems for veterinary vaccines have recently been reviewed by Gerdts et al. (Gerdts et al., 2013).

Recent progress in particle-based delivery of veterinary vaccines is mostly limited to experiments in murine models. For example, solid nanolipid particles with inactivated FMDV and IL-2 (Wang et al., 2016), quaternized chitosan hydrogel microparticles (Gel MPs) with inactivated PRRSV (Chen et al., 2014), and (hollow) silica vesicles with BVDV E2 protein (Mody et al., 2015; Mody et al., 2014) and *Anaplasma marginale* (Zhang et al., 2016a) VirB9-1 and VirB10 were generated, tested in mice, and shown to elicit a strong immune response; however, the efficacy of these particle-based vaccine formulations needs to be confirmed in the target species.

A new hollow-type mesoporous silica nanoparticle with surface amino functionalization (HMSA) was formulated with BVDV E2 protein, freeze-dried, and tested in sheep. After three subcutaneous (SC) immunizations with the E2/HMSA, sheep developed antibody levels equivalent to those induced with E2 formulated with Quil A, as well as robust CMI responses. The fact that the E2/HMSA retained activity after freeze-drying in sheep makes it a practical attractive vaccine candidate (Mahony et al., 2015). In another study, BVDV E2 and NS3 and poly(I:C) were formulated in PLGA as a biomimetic with E2 on the surface and NS3 and poly(I:C) inside the particles. After two SC immunizations, the Vaccine-NP elicited NS3- and E2- specific antibodies with VN activity, as well as IFN-γ responses in PBMCs stimulated *ex vivo* with E2 and NS3 proteins in sheep. These responses were equivalent to those elicited by an inactivated BVDV vaccine, and protective against BVDV-1 challenge (Riitho et al., 2017).

Chitosan and silica nanoparticles are also promising in chickens as delivery vehicle for avian leukosis virus (ALV) gp85. The antibody responses were longer lasting and the viremia after ALV challenge shorter when gp85 was formulated with silica nanoparticles than in Freund's adjuvant (Cheng et al., 2017). Intranasal delivery of NDV and IBV antigens formulated with quaternized chitosan nanoparticles were shown to elicit higher IgG, IgA, and lymphocyte proliferation than a commercial attenuated NDV/IBV vaccine in chickens, as well as protection from NDV and IBV challenge (Zhao et al., 2017).

Liposomes and virosomes

Liposomes have been described for many years, since 1940. They consist of vesicles made up of lipid layers, with antigens or plasmid DNA, and sometimes other adjuvant components incorporated. The properties of liposomes depend on the type of lipids and surfactants employed, which determine the size, polarity, charge and number of layers (uni- or multilamellar). Liposomes primarily function as carriers facilitating uptake of antigens by APCs, but can also have immunostimulatory properties. Cationic liposomes usually are more readily taken up by APCs than anionic liposomes. The use of cationic liposomes in animal models has been reviewed by Korsholm et al., including some examples of uses in pigs for PRV, cattle for BVDV and bovine tuberculosis (TB), sheep for bovine leukemia virus and chickens for NDV (Korsholm et al., 2012).

More recently, a liposome was tested as adjuvant for a porcine circovirus (PCV)2 DNA vaccine. Liposomes were prepared from 1,2-dimyristoyl-*sn*-glycero-3-phosphoethanolamine, dihexadecyl phosphate, and cholesterol. PCV2 ORF2 DNA was formulated with the liposomes, cobalt oxide or aluminum hydroxide. Pigs immunized with the liposome-formulated DNA vaccine had significantly higher VN antibody levels and numbers of IFN-γ-secreting cells, as well as lower PCV2 viremia, than the pigs immunized with the other experimental vaccines. The pigs immunized with the DNA formulated with liposome or cobalt oxide had lower lymphoid lesion scores than the pigs inoculated with the PCV2 DNA vaccine in PBS (Park et al., 2017). The large-scale production of liposomes can be a limiting factor, and liposomes are often not very stable. Overall, liposomes have been rarely tested in veterinary species. However, CAF(01) is a stable liposome-based adjuvant discussed under combination adjuvants that has been tested in human clinical trials, as well as with *Mycobacterium bovis* and *M. avium* in cattle.

Virosomes are unilamellar lipid vesicles prepared from enveloped viruses, usually influenza virus, which contain the viral glycoproteins, and can be used to incorporate various molecules such as proteins, peptides, and cytokines, which then are displayed on the virosomal surface. In contrast to liposomes, virosomes, due to the presence of viral attachment and fusion proteins such as the infuenza virus HA, can fuse with host target cells. Virosomes are being explored for delivery of vaccine antigens, nucleic acids, drugs, and antibodies (Gerdts et al., 2013). However, like liposomes, virosomes are infrequently evaluated for vaccine delivery in large animal species.

Combination adjuvants

Many highly effective adjuvants contain more than one active compound and are termed combination adjuvants. For example, the well-known Freund's complete adjuvant is a water-in-oil emulsion with mycobacteria and the less toxic RIbi adjuvant (Vaxio) contains MPL and cell wall skeleton in squalene. Combination adjuvants may offer several advantages: i) combination of various adjuvant functions, for example depot and immunostimulation; ii) a synergistic action leading to promotion of a stronger and/or broader immune responses; and iii) reduction of individual components to reduce toxicity. Many combination adjuvants in development contain a TLR agonist, which allows for tailoring the bias of the immune response, and a delivery vehicle. A number of new combination adjuvants have recently been developed and commercialized, including MF59 (Novartis Inc.), Adjuvant Systems™ (ASO) (Glaxo Smith Kline Inc.) and IC31 (Valneva Inc.). One of the adjuvant systems, AS03, is part of licensed human vaccines against HPV and HBV. We recently reviewed vaccine potentiation by combination adjuvants including pre-clinical and clinical studies in context of human and animal vaccines (Levast et al., 2014).

Dimethyldioctadenylammonium (DDA), an adjuvant used since 1966, is not very powerful, but when combined with other compounds, can elicit robust immune responses. For example, we showed that gD in combination with DDA and Emulsigen® elicits robust immune responses and protection from BoHV-1 (Van Donkersgoed et al., 1996), The adjuvant CAF01 contains DDA and trehalose dibehenate (TDB); TDB stabilizes the liposomes and stimulates T cell responses to *Mycobacterium bovis* Ag85B-ESAT-6 (Davidsen et al., 2005). Other adjuvants co-formulated with DDA, MPL and monomycoloyl glycerol (MMG), also promoted T cell responses, but CAF01 was moved forward into clinical trials (Levast et al., 2014). Recently, addition of poly(I:C) to either TDB or MMG, which both target the monocyte-inducible C-type lectin (MINCLE) receptor, enhanced bovine monocyte activation. CAF09, which contains MMG and poly(I:C), was formulated with *Mycobacterium avium* proteins and found to elicit enhanced humoral and CD4 T cell responses in vaccinated calves when compared to CAF01, including a higher frequency of polyfunctional T cells (Thakur et al., 2018). However, CAF09 forms a depot at the injection site when delivered IM or SC, and does not elicit CTL responses. Recently, a series of nanoemulsions (CAF24a-c) was designed based on CAF09, as an approach to promote CTL responses. The oil phase consisted of biodegradable squalene, and DDA was replaced by zwitterionic distearoyl-phosphoethanolamine to vary the surface charge. CAF24a-c reached the draining LN, and mediated cross-presentation by DCs and induction of antigen-specific CD8 CTL responses. This makes CAF24a-c promising for induction of CTL responses upon SC and IM immunization, which is important for vaccines against pathogens for which CTL responses are required to prevent infection (Schmidt et al., 2018).

Several saponin adjuvant combinations have been evaluated in cattle. Saponin and Emulsigen® were formulated as a combination with *Mycoplasma bovis* cells, and injected SC into calves. Vaccinated calves developed strong humoral immune responses, and after challenge, gross pathological lung lesions and nasal shedding of *M. bovis* were reduced (Dudek et al., 2016). *Tritrichonomas foetus* cells were formulated with saponin, alum or both adjuvants, and administered SC to heifers. The IgG and IgA levels were increased and the clearance of *T. foetus* was most rapid in animals vaccinated with the combination adjuvant (Fuchs et al., 2017). In the context of *Staphylococcus aureus*, the α-toxoid was formulated with a proprietary Alum-based adjuvant (Brenntag) supplemented with either saponin (QuilA; Brenntag), light mineral oil, or both, and administered SC to cows. The α-toxoid in Alum–Saponin–Oil elicited higher specific antibody titers in milk and serum compared to α-toxoid in Alum–Oil and Alum–Saponin (Boerhout et al., 2018).

In chickens, multiple adjuvants have been evaluated for their synergistic activity. For example, a number of agonists of membrane and cytosolic TLRs were evaluated individually and in combination *ex vivo*. Bone marrow derived macrophages of 1- and 4-week old birds were stimulated with Pam3Cys-SK4 (PCSK; TLR2), synthetic MPLA, Di[3-deoxy-D-manno-octulosonyl]-lipid A ammonium salt (KLA; TLR4), Gardiquimod, Resiquimod (R848; TLR7), or CpG class B and C (TLR21). Nitric oxide (NO) production and mRNA levels of IL-1β, IL-10 and IL-12p40 were higher in the stimulated macrophages from 4-week-old birds compared to 1-week-old birds. Co-stimulation by CpG B and MPLA or PCSK resulted in enhanced NO produciton and inflammatory gene expression (Kim et al., 2018). Recently, a ginseng stem-leaf saponin (GSLS) and selenium (Se) were shown to have a synergistic adjuvant effect in chickens *in vivo*. A live NDV/IBV vaccine was formulated with GSLS, Se or both, and administered via the ocular or intranasal route to two day-old chickens. The chickens that received the NDV/IBV vaccine with both GSLS and SE developed the highest VN antibody responses, lymphocyte proliferation and IFN-γ and IL-4 production. These responses also developed earlier and lasted longer (Ma et al., 2019).

We recently developed a combination adjuvant (TriAdj) that consists of either CpG ODN or poly(I:C), a immune defense regulatory (IDR) peptide and poly[di(sodium carboxylatoethylphenoxy)]-phosphazene (PCEP), which has applications for both human and veterinary vaccines (reviewed by Garg et al. (Garg et al., 2017)). CpG ODN and poly(I:C) induce the production of pro-inflammatory cytokines, activate DCs, macrophages, natural killer cells and B cells, and promote Th1/ balanced immune responses, including CD8 T cells (Tewari et al., 2010). The IDR peptide is a derivative of a host defense peptide, and has antimicrobial, chemotactic and immmunomodulatory properties (Yeung et al., 2011). Polyphosphazene is a synthetic polymer that also has immunostimulatory properties and acts as a delivery vehicle by forming non-covalent complexes with the antigen and adjuvant components (Awate et al., 2012). The components in TriAdj act synergistically and are needed for optimal efficacy (Kovacs-Nolan et al., 2009a). In the context of veterinary vaccines, the TriAdj has strong immunostimulatory properties when used to adjuvant antigens from bovine respiratory syncytial virus (BRSV), BVDV, porcine epidemic diarrhea virus (PEDV) and C*hlamydia pecorum* in multiple species including mice, cotton rats, sheep, pigs, cattle, and koalas (Kovacs-Nolan et al., 2009b) (Desclozeaux et al., 2017; Garg et al., 2015; Garg et al., 2014; Khan et al., 2014; Snider et al., 2014). For example, when formulated with low doses (25 – 50 µg) of BVDV E2 protein, TriAdj mediated induction of robust humoral and CMI responses after two IM or ID immunizations, resulting in protection from virulent type 2 BVDV challenge (Sadat et al., 2017; Snider et al., 2014). The PEDV spike (S1) protein was formulated with TriAdj, and three consecutive doses were administered IM to pregnant sows. Increased S1-specific IgG, IgA and VN antibody levels were induced in the sera and colostrum of the vaccinated sows. The piglets born to the vaccinated sows had high serum IgG titers, and 87% of them was protected from PEDV challenge (Makadiya et al., 2016).The chlamydial major outer membrane protein antigen was formulated with TriAdj, and shown to induce long lasting (~1 year) humoral and CMI responses, including neutralizing antibodies, in female koalas after a single immunization. *C. pecorum*- specific neutralizing antibodies were found in the plasma as well as vaginal and conjunctival secretions (Khan et al., 2014).

Mechanisms of action of adjuvants

Adjuvants have been used in billions of vaccine doses for humans and animals, but until recently, relatively little was known about how they modulate antigen-specific immune responses. In the last decade there has been a great deal of interest in the mechanisms of action (MOA) of adjuvants. Several MOAs have been revealed, mostly based on investigations in laboratory animals; there are numerous reviews in the literature and the reader is referred to such publications for details. There are relatively few studies in veterinary species, and these studies will be highlighted. While there still is much to be learned about their MOAs, the general consensus is that adjuvants mediate their effects through processes that entail one or more of the following events: sustained release of antigen from injection site (depot effect), modulation of cytokines and chemokines, cell recruitment, increased antigen uptake and presentation, activation and maturation of APCs, migration of cells to the draining LNs, and activation of cellular mechanisms in immune cells (Awate et al., 2013). A single adjuvant may have several MOAs, and it is thought that adjuvant modulation of DC activation, maturation and recruitment, as well as antigen processing and presentation are critical in this process (Duthie et al., 2011; Reed et al., 2009).

Depot formation

For a long time, depot formation was the most recognized mechanism by which adjuvants enhance antigen-specific immune responses. Typically, the antigen and adjuvant are co-mixed prior to injection into the host. The antigen is then 'trapped' in the adjuvant and retained at the injection site and subsequently slowly released over a prolonged period. This is thought to contribute to sustained stimulation of the immune system resulting in stronger immune responses. Depot formation was first proposed in 1926 by Glenny (Glenny, 1926) and has since been considered an importance mechanism by which Alum enhances antibody responses to antigens; however, numerous studies have suggested otherwise. Whether depot formation significantly contributes to the adjuvant activity of Alum remains controversial (Marrack et al., 2009).

More recently, Didielaurent and colleagues explored whether depot formation contributed to the adjuvant activity of AS04, a combination adjuvant consisting of MPL, a LPS derivative, and Alum, which is a component in human licensed vaccines (Didierlaurent et al., 2009). They reported that in mice, AS04 was most effective when co-injected with antigen, but not when injected at separate sites (Didierlaurent et al., 2009). The authors concluded that perhaps Alum played a role in stabilizing the MPL and antigen, along with providing the depot effect. While formation of depot may contribute to adjuvant activity, it remains controversial whether it is required for adjuvant activity. In fact, it has been argued that more evidence is required to prove that depot formation plays any significant role in adjuvant activity (Marrack et al., 2009). It has been reported that a number of adjuvants, including MF59 and ISCOMS, do not form a depot as they are rapidly (within hours) cleared from the injection site, suggesting that at least for these adjuvants and many others, depot formation may not play a significant role in their activity.

It is important to note that depot is used almost exclusively in reference to injected vaccines and all the studies on depot effect were performed with injectable adjuvants. There are no published reports on investigations regarding the contribution of depot formation in mucosally administered adjuvants. It is reasonable to speculate that the role of depot formation in mucosally administered vaccines is most likely minimal at best, and almost certainly not feasible in orally administered vaccines. To our knowledge all published reports on depot formation have been performed in laboratory animals, and mostly in mice. Large animals present a more robust model to investigate MOAs of adjuvants because the large body size allows access to a variety of tissue samples; however, this also makes it difficult to localize the injection site, specifically in case of IM immunization.

Regulation of immune response genes

Most adjuvants, with a few exceptions, are substances that are 'foreign' to the host. As such they are expected to elicit a reaction at the site of injection as a normal host response to any insult. Following recognition of the adjuvant by the immune system, the first indication of a host response is the modulation (upregulation or downregulation) of genes involved in innate immune activation at the site of injection. Mosca and colleagues coined the term "adjuvant core response genes", a set of 56 genes modulated by the three adjuvants CpG ODN, alum and MF59 (Mosca et al., 2008). These adjuvant core response genes include cytokines, chemokines, innate immune receptors, IFN-induced genes and adhesion molecules (Mosca et al., 2008). In studies in mice, the experimental adjuvant PCEP induced time-dependent changes in expression of a variety of genes including cytokine, chemokine, innate immune receptor, IFN-induced, adhesion molecule and antigen presentation genes (Awate et al., 2012). In addition, PCEP increased the expression of the gene encoding the inflammasome receptor, NLRP3, suggesting that it may be involved in activation of the inflammasome, which is required for the release of IL-1 and promotes inflammatory responses (Awate et al., 2012).

In subsequent similar studies in pigs, it was observed that PCEP induces changes in gene expression of many molecules. Specifically, PCEP induced increased expression of genes encoding the chemokine CCL-2 and the proinflammatory cytokine IL-6 (Magiri et al., 2016). In contrast, CpG DNA induced increased gene expression of the chemokine CXCL10, several IFN response genes, TNF-α, and IL-10, and downregulation of IL-17 genes, while Emulsigen® induced increased gene expression of CCL2 and CCL5, IL-6 and TNF-α, TLR9 and several INF response genes (Magiri et al., 2016). Quite clearly, these three adjuvants had remarkably different effects on innate immune response-related genes, and this may explain why their adjuvant activity differs. Furthermore, the effects of PCEP on immune response genes differed between mice and pigs, which confirmed that adjuvant activity can differ between species. This observation strongly suggests that observations from studies in mice should not be directly extrapolated to large animals.

The adjuvant combination TriAdj, which contains PCEP, was studied at the level of immune response genes and proteins, as well as cell signaling. When administered intranasally, the RSV fusion protein formulated with TriAdj induced production of monocyte-recruiting chemokines, CCL2, CCL3 and CCL7, and DC-recruiting chemokines, CCL3, CCL4 and CXCL-10, in both nasal-associated tissue (NALT) and lung. In addition, the pro-inflammatory cytokines TNF-α, IL-1β and IL-6 were produced in response to immunization with the TriAdj-formulated fusion protein in both tissues (Sarkar et al., 2016). In macrophages, which was a major recruited cell population, JNK, ERK1/2, CaMKII, PI3K and JAK pathways were responsible for the chemokine and pro-inflammatory cytokine responses mediated by the TriAdj-formulated fusion protein, while the p38 and NF-κB pathways appeared to be not or minimally involved (Sarkar et al., 2018). Intradermal immunization of calves with the BVDV E2 protein formulated with TriAdj resulted in local gene expression of IL-1β, IL-6, CXCL8 and CCL2 (Sadat et al., 2017).

Regulation of immunomodulatory proteins

As discussed above, adjuvants induce changes in gene expression of many immune-related genes. However, not all genes are translated into proteins. In this regard, investigators have typically followed gene expression studies with assessment at the protein level to determine changes in proteins involved in mediating immunity including cytokines and chemokines (Awate et al., 2013; Awate et al., 2012), which influence certain events related to adjuvant activity such as cell recruitment, activation, and trafficking to the draining LNs. For example the TriAdj formulation mediated production of CCL2, CXCL1, TNF-α, IL-6, IL-β, IFN-γ, IL-12p70 and IL-10 at the protein level (Sarkar et al., 2016).

In contrast to the numerous reports in mice on how adjuvants modulate cytokine and chemokine production, there are very few systematic investigations on this aspect in large animals. In recent studies, pigs were injected with varying doses of the adjuvant PCEP followed by an assessment of a variety of cytokines and chemokines in the serum and locally at the site of injection. PCEP induced increased production of IL-1β and IL-13 at the site of injection (skin) and IL-1β and IL-6 in the draining LNs (Magiri et al., 2016). Emulsigen® induced increased production of IL-1β, IL-6, and IL-12 in the skin, but not in the draining LNs (Magiri et al., 2016). No cytokines or chemokines were detected in the serum, suggesting that neither of these adjuvants had a systemic effects, and thus, that they are relatively safe (Magiri et al., 2016).

Cell recruitment

As many practicing veterinarians may be aware, some vaccines induce inflammation, which may be manifested as a painful swelling at the site of injection, or transient fever that may lead to loss of appetite or decreased production. It is generally thought that a certain amount of inflammation is necessary for a vaccine to be effective. Inflammation is a hallmark of innate immunity. One of the key features of inflammation is cell recruitment to a site of either infection or any other injury to the tissue that provokes a host to mount an immune response. There are numerous studies in mice that have evaluated local cell recruitment in response to various adjuvants. The capacity of MF59 to induce recruitment of cells has been studied in detail (Calabro et al., 2011). At the site of injection, this adjuvant induced recruitment of neutrophils, monocytes, eosinophils, macrophages, DCs and B cells (Calabro et al., 2011). However, depletion of neutrophils had no effect on adjuvant activity (Calabro et al., 2011), implying that the role of neutrophils in adjuvant activity remains unresolved. Similarly, AS03 induced enhanced local recruitment of neutrophils, eosinophils and monocytes.

Intramuscular injection of PCEP in mice induced recruitment of neutrophils, macrophages, monocytes, DCs and also lymphocytes including CD4 and CD8 T cells, as well as B cells, to the site of injection and in the draining LNs (Awate et al., 2014a; Awate et al., 2014b). Intradermal injection of PCEP in pigs promoted infiltration of macrophages, lymphocytes and neutrophils in the skin 14 days after injection (Magiri et al., 2019). Emulsigen® induced skin infiltration with mostly macrophages and lymphocytes, and small granulomas were seen (Magiri et al., 2019). In addition to cell recruitment, necrotic debris were observed, suggesting cell death (Magiri et al., 2019). Cell death by necrosis is accompanied by release of endogenous cytosolic molecules, some of which have been associated with adjuvant activity as they activate DAMPs, resulting in activation of innate immunity.

Subcutaneous injection of CpG ODN in sheep induced cellular infiltration of IFN-α and IFN-γ producing cells, most of which were primarily CD172 myeloid cells (Nichani et al., 2004). The presence of IFN-α and IFN-γ in the skin was CpG ODN-induced and confirmed that CpG ODN not only recruited cells, but also activated them locally (Nichani et al., 2004).

Intranasal administration of the RSV fusion protein with TriAdj resulted in recruitment of macrophages, DCs and neutrophils into the NALT and lung, which correlated to enhanced expression of CCL2, CXCL1, and pro-inflammatory cytokines (Sarkar et al., 2016). TriAdj, when formulated with the BVDV E2 protein, also mediated recruitment of DCs, monocytes, macrophages and neutrophils into the skin of ID vaccinated calves (Sadat et al., 2017).

Antigen uptake

Antigen presenting cells more readily take up particulates than soluble proteins. Consequently, several types of adjuvants, including alum, liposomes, synthetic polymers and micro/nanoparticles, act at least partially by enhancing uptake of the antigen by APCs, i.e. DCs, macrophages and B cells. Micro- and nanoparticles and alum aggregates can be more easily phagocytosed. Carbohydrates can bind to receptors on APCs, and liposomes can deliver the antigen by fusing with the cellular membrane. As an example, the TriAdj, which contains poly(I:C), a cationic peptide and PCEP, mediated enhanced uptake of the RSV fusion protein by macrophages

(Sarkar et al., 2018), which can be attributed to the formation of nanoparticles by the PCEP.

Recently, a new type of adjuvant named Nano-11 was developed, and evaluated in piglets. Nano-11 nanoparticles were formed from phytoglycogen, a dendromer derived from corn. Nano-11 significantly improved the systemic and mucosal antibody responses to killed influenza virus antigen induced by intranasal administration in piglets, which correlated to enhanced protection from viral challenge. One of the MOAs of Nano-11 was enhanced uptake by porcine monocyte-derived DCs, in addition to secretion of pro-inflammatory cytokines (Dhakal et al., 2019).

In a number of studies in sheep afferent lymph was analyzed for the presence of immune cell populations that take up antigen. Formulation with aluminum was found to significantly enhance uptake of particulate antigen by DCs from 48 to 72 h after antigen injection (de Veer et al., 2010). Interestingly, the amount of antigen per cell was enhanced by the addition of poly(I:C) into the formulation (Neeland et al., 2014a). In another study, antigen formulated with liposomes was found to be taken up by migratory DCs, and the presence of CpG ODN in the formulation increased the number of DCs and neutrophils carrying antigen. CpG ODN and poly(I:C) both increased the systemic antigen-specific humoral immune responses in sheep (Neeland et al., 2014b).

Cellular activation

Most modern adjuvants that are currently under development enhance antigen-specific immune responses by activating innate immunity through engagement of one or more of the PRRs. Thus, these adjuvants are natural or synthetic agonists for PRRS. A few of these receptors and their agonists have received great attention and will be highlighted below.

TLR9 agonists: The natural agonist for TLR9 is microbial DNA containing CpG DNA. However, synthetic CpG ODNs are the most commonly used agonists for TLR9 activation. CpG ODN directly activates mainly plasmacytoid DCs to secrete IFN-α, and indirectly activates many other cell types to secrete a number of cytokines that promote Th1-type immune responses. Several factors are thought to contribute to the adjuvant activity of CpG ODN including synergistic activation of TLR9 and B-cell receptors, inhibition of B cell apoptosis, and DC maturation resulting in activation of Th1 cells and strong CTL responses. CpG ODNs, either alone or in combination with other adjuvants, have been experimentally used as adjuvants in several veterinary species, as has been discussed earlier in this chapter.

TLR4 agonists: The natural agonist for TLR4 is LPS, but this molecule is highly toxic and as such has limited application as an adjuvant due to safety concerns. Less toxic forms of LPS including MPL and glucopyranosyl lipid A (GLA) have been developed. Both MPL and GLA, either alone or in combination with other adjuvants, have been tested extensively in mice and to a lesser extent in humans, but rarely in large animals. GLA-SE, a combination adjuvant containing GLA and squalene (an oil-in-water emulsion) increased IgG1 antibodies specific to ID83 antigen of *Mycobacterium bovis* in cattle (Jones et al., 2013). GLA-SE did not have any effect on antigen-specific IFN-γ production or cell proliferation (Jones et al., 2013), suggesting that this adjuvant promotes a Th2-, but not Th1-type immune response, the latter of which is critical for protection against mycobacterial infections. Minipigs immunized with ESAT-6 (a recombinant *Mycobacterium tuberculosis* antigen) formulated with a combination adjuvant containing GLA had increased antibody responses after ID but not intranasal immunization (McKay et al., 2016). This observation strongly suggests that the route of administration of an adjuvant is critical, and may influence the MOA.

MINCLE agonists: It has been recognized for a long time that mycobacterial components have strong adjuvant activity and this is clearly evident from the decades of use of Complete Freund's adjuvant in laboratory animals. Trehalose demycolate (TDM) is a component in the mycobacterial cell wall that is associated with adjuvant activity. However, until recently the receptor for TDM was unknown. In 2009, MINCLE, a membrane-associated C-type lectin receptor, was identified as the receptor for TDM (Ishikawa et al., 2009). A year later, the same receptor was associated with cellular activation by TDM's synthetic analog, TDB (Schoenen et al., 2010). Furthermore, it has been established that activation of the MINCLE-dependent pathway is required for the adjuvant activity of TDM and TDB, which are primarily associated with promotion of combined Th1 and Th17-type immune responses (Schoenen et al., 2010; Werninghaus et al., 2009). In this regard,

MINCLE agonists are prime candidates to be tested with vaccines where cellular immunity is required. For example, GlcC14C18, a synthetic MINCLE agonist that is less toxic than TDM and TDB, was found to induce protective immunity in a mouse model of *Mycobacterium tuberculosis*, when incorporated in a cationic liposome formulation (Decout et al., 2017). To date, no studies have evaluated MINCLE agonists in veterinary species.

Antigen processing and presentation

Once taken up by the APCs, such as DCs, macrophages and B cells, antigens are processed and presented to T cells on MHC molecules. Adjuvants can influence the efficiency of antigen processing and presentation by increasing MHC expression and the level of co-stimulatory molecules on APCs. Generally, exogenously acquired antigens are presented on MHC class II molecules leading to a Th2-type, primarily humoral, response, while endogenously processed antigens are presented on MHC class I molecules resulting a in a Th1-type, mostly cell-mediated, immune response. Some adjuvants, for example liposomes and poly(I:C), can promote cross-presentation of exogenous antigens and CD8 T cell responses. In addition, depending on the danger signal effected, adjuvants can enhance the production of cytokines by APCs that further influence the type of immune response, i.e. Th1 cytokines IL-12, IFN-γ and IL-2, Th2 cytokines IL-4, IL-5 and IL-13 or cytokines that lead to a Th17 or regulatory T cell response.

As for the majority of the studies on adjuvant MOI, most information is based on murine models. For example, we have shown enhanced expression of CD86, MHC-II and VCAM on DCs and macrophages in the NALT and lungs of mice immunized with a TriAdj – formulated subunit vaccine (Sarkar et al., 2016). In a recent study a new adjuvant, resveratrol, was formulated with a PRV vaccine, and shown to enhance antigen presentation of peritoneal macrophages by upregulating the expression of MHC-II and CD86 (Chen et al., 2019). In another *ex vivo* study the immunostimulatory activity of the TLR2 ligand Pam3Cys and the TLR9 ligand CpG ODN were compared for neonatal and adult porcine DCs. Pam3Cys induced more TNF-α production and CD80/86 upregulation in monocytes and plasmacytoid DCs from neonates than from adult pigs. In addition, CpG ODN promoted more IL-12 mRNA expression in neonatal mononuclear cells than in those from adults (Vreman et al., 2018).

In an *in vivo* study, pigs were immunized twice intranasally with Norovirus P particle containing SwIV M2e protein and two H1N1 peptides from pandemic and classical influenza virus and formulated in PLGA nanoparticles. Subsequently they were challenged with heterologous SwIV H1N1. While the antibody levels were not increased, the numbers of both CD4 and CD8 antigen-specific, IFN-γ antigen-specific secreting T cells were increased in the lungs of the vaccinated pigs, demonstrating promotion of cross-presentation by the PLGA nanoparticles. In addition, the vaccinated pigs did not develop fever or clinical symptoms (Hiremath et al., 2016).

The effects of another adjuvant consisting of bacterial ghosts (BGs) on bovine monocyte-derived DCs (Mo-DCs) was evaluated *ex vivo*. The BGs induced earlier and higher expression of MHC-II, CD80 and CD86, as well as higher levels of IL-12 and IL-10, in BG-stimulated Mo-DCs than in LPS-stimulated cells. In addition, the BG-stimulated Mo-DCs stimulated higher allogeneic T cell proliferation in a mixed lymphocyte reaction, demonstrating maturation of Mo-DCs (Hajam et al., 2015).

Effects on draining lymph nodes

There is ample evidence that parenteral injection of adjuvants results in significant immunological changes in the draining LNs. It is thought that injection of adjuvant induces local production of chemokines and cytokines that mediate immune cell recruitment to the injection site, which then traffick to the draining LNs. In this regard, injection of the adjuvants MF59 and AS03 in mice induced local recruitment of neutrophils, monocytes, macrophages and DCs which then traffick to the draining LNs (Calabro et al., 2011; Morel et al., 2011). In mice, PCEP induced a significant increase in immune cells present in draining LNs including neutrophils, macrophages, monocytes, DCs and also T and B lymphocytes (Awate et al., 2014a; Awate et al., 2014b).

In pigs, injection of PCEP induced significant production of IL-1β and IL-6 (Magiri et al., 2016), while Emulsigen® did not induce any significant cytokine or chemokine production in the draining LNs (Magiri et al., 2016). In addition, PCEP and Emulsigen® induced LN infiltration of macrophages, monocytes, T and B cells, as well as necrotic debris in pigs (Magiri et al., 2019).

Conclusions and future perspectives

For a long time, Alum and oil-based adjuvanst were the only adjuvants used in animals; however, in the last 20 years, six adjuvants were approved, so a few veterinary and human vaccines contain new adjuvants. Overall though, progress has been slow as adjuvants need to be registered as an entity of a combination with the antigen. At the pre-clinical level, many new adjuvant types are under investigation, and although much research still needs to be done, we have increased knowledge of the MOA of adjuvants; however, while some of the MOAs have been investigated in veterinary species as described in this chapter, the majority of them has been studied in mice. Information about adjuvants in registered, commercial vaccines is frequently proprietary. This also applies to some new-generation adjuvants, which makes it difficult to test them in independent trials. Nevertheless, we now have a better rationale for selection of adjuvants suitable for specific antigens, diseases, animal species and routes of delivery. For example, in order to elicit protective immunity to extracellular pathogens, humoral immune responses can be sufficient, while CMI responses, often in addition to antibody responses, are critical to protect against intracellular pathogens. We have a much better understanding of the innate immune system, the link with adaptive immunity, and how adjuvants can activate PRRs, which mediate activation of the innate response. Since multiple receptors and ligands are involved in activation of the innate immune response, adjuvant formulations can be generated that contain one or several of those ligands in combinations that are tailor-made for induction of balanced Th1/Th2, Th1 or Th2 responses. Indeed many new adjuvants combine several MOAs that can be additive or even synergistic.

While a certain level of inflammation is correlated to the induction of an immune response to vaccination, a balance must be maintained between the initiation of the immune response and the level of systemic and local reactions. Our ability to evaluate the immune responses at multiple levels including activation of the innate immune system through analyses of cell recruitment and activation, gene and protein expression of signaling molecules and metabolite changes, will allow us to do an in-depth characterization of the MOA of new adjuvants. This will provide critical information about both the efficacy and safety of new adjuvants, facilitate optimization for different pathogens, species, and routes of delivery, as well expedite licensing as part of new vaccines. In the context of veterinary vaccines, more so than for human vaccines, economics play a major role. Potent adjuvants will allow us to reduce the amount of antigen needed per dose (antigen sparing) and possibly reduce the number of vaccinations needed, and thus develop new killed and subunit vaccines that are safe, effective and affordable.

References

Adair, B.M., 2009. Nanoparticle vaccines against respiratory viruses. Wiley interdisciplinary reviews 1, 405-414.

Ahlberg, V., Hjertner, B., Wallgren, P., Hellman, S., Lovgren Bengtsson, K., Fossum, C., 2017. Innate immune responses induced by the saponin adjuvant Matrix-M in specific pathogen free pigs. Vet Res 48, 30.

Angelos, J.A., Edman, J.M., Chigerwe, M., 2014. Ocular immune responses in steers following intranasal vaccination with recombinant Moraxella bovis cytotoxin adjuvanted with polyacrylic acid. Clin Vaccine Immunol 21, 181-187.

Aoshi, T., 2017. Modes of Action for Mucosal Vaccine Adjuvants. Viral immunology 30, 463-470.

Awate, S., Babiuk, L.A., Mutwiri, G., 2013. Mechanisms of action of adjuvants. Frontiers in immunology 4, 114.

Awate, S., Eng, N.F., Gerdts, V., Babiuk, L.A., Mutwiri, G., 2014a. Caspase-1 Dependent IL-1beta Secretion and Antigen-Specific T-Cell Activation by the Novel Adjuvant, PCEP. Vaccines 2, 500-514.

Awate, S., Wilson, H.L., Lai, K., Babiuk, L.A., Mutwiri, G., 2012. Activation of adjuvant core response genes by the novel adjuvant PCEP. Molecular immunology 51, 292-303.

Awate, S., Wilson, H.L., Singh, B., Babiuk, L.A., Mutwiri, G., 2014b. The adjuvant PCEP induces recruitment of myeloid and lymphoid cells at the injection site and draining lymph node. Vaccine 32, 2420-2427.

Blome, S., Wernike, K., Reimann, I., Konig, P., Moss, C., Beer, M., 2017. A decade of research into classical swine fever marker vaccine CP7_E2alf (Suvaxyn((R)) CSF Marker): a review of vaccine properties. Vet Res 48, 51.

Boerhout, E.M., Koets, A.P., Mols-Vorstermans, T.G.T., Nuijten, P.J.M., Hoeijmakers, M.J.H., Rutten, V., Bijlsma, J.J.E., 2018. The antibody response in the bovine mammary gland is influenced by the adjuvant and the site of subcutaneous vaccination. Vet Res 49, 25.

Borrego, B., Blanco, E., Rodriguez Pulido, M., Mateos, F., Lorenzo, G., Cardillo, S., Smitsaart, E., Sobrino, F., Saiz, M., 2017. Combined administration of synthetic RNA and a conventional vaccine improves immune responses and protection against foot-and-mouth disease virus in swine. Antiviral research 142, 30-36.

Calabro, S., Tortoli, M., Baudner, B.C., Pacitto, A., Cortese, M., O'Hagan, D.T., De Gregorio, E., Seubert, A., Wack, A., 2011. Vaccine adjuvants alum and MF59 induce rapid recruitment of neutrophils and monocytes that participate in antigen transport to draining lymph nodes. Vaccine 29, 1812-1823.

Camussone, C.M., Pujato, N., Renna, M.S., Veaute, C.M., Morein, B., Marcipar, I.S., Calvinho, L.F., 2014a. Immune response and functional role of antibodies raised in heifers against a Staphylococcus aureus CP5 lysate and recombinant antigens vaccine formulated with Iscom Matrix adjuvant. Veterinary immunology and immunopathology 162, 96-107.

Camussone, C.M., Veaute, C.M., Pujato, N., Morein, B., Marcipar, I.S., Calvinho, L.F., 2014b. Immune response of heifers against a Staphylococcus aureus CP5 whole cell and lysate vaccine formulated with ISCOM Matrix adjuvant. Res Vet Sci 96, 86-94.

Chambers, M.A., Graham, S.P., La Ragione, R.M., 2016. Challenges in Veterinary Vaccine Development and Immunization. Methods Mol Biol 1404, 3-35.

Chen, G., Zeng, S., Jia, H., He, X., Fang, Y., Jing, Z., Cai, X., 2014. Adjuvant effect enhancement of porcine interleukin-2 packaged into solid lipid nanoparticles. Res Vet Sci 96, 62-68.

Chen, M., Chen, X., Song, X., Muhammad, A., Jia, R., Zou, Y., Yin, L., Li, L., He, C., Ye, G., Lv, C., Zhang, W., Yin, Z., 2019. The immune-adjuvant activity and the mechanism of resveratrol on pseudorabies virus vaccine in a mouse model. Int Immunopharmacol 76, 105876.

Cheng, J., Wen, S., Wang, S., Hao, P., Cheng, Z., Liu, Y., Zhao, P., Liu, J., 2017. gp85 protein vaccine adjuvanted with silica nanoparticles against ALV-J in chickens. Vaccine 35, 293-298.

Christensen, M.H., Paludan, S.R., 2017. Viral evasion of DNA-stimulated innate immune responses. Cell Mol Immunol 14, 4-13.

Dalsgaard, 1978. A study of the isolation and characterization of the saponin Quil A. Evaluation of its adjuvant activity, with a special reference to the application in the vaccination of cattle against foot-and-mouth disease. Acta Vet Scand Suppl, 7-40.

Davidsen, J., Rosenkrands, I., Christensen, D., Vangala, A., Kirby, D., Perrie, Y., Agger, E.M., Andersen, P., 2005. Characterization of cationic liposomes based on dimethyldioctadecylammonium and synthetic cord factor from M. tuberculosis (trehalose 6,6'-dibehenate)-a novel adjuvant inducing both strong CMI and antibody responses. Biochimica et biophysica acta 1718, 22-31.

de Veer, M., Kemp, J., Chatelier, J., Elhay, M.J., Meeusen, E.N., 2010. The kinetics of soluble and particulate antigen trafficking in the afferent lymph, and its modulation by aluminum-based adjuvant. Vaccine 28, 6597-6602.

Decout, A., Silva-Gomes, S., Drocourt, D., Barbe, S., Andre, I., Cueto, F.J., Lioux, T., Sancho, D., Perouzel, E., Vercellone, A., Prandi, J., Gilleron, M., Tiraby, G., Nigou, J., 2017. Rational design of adjuvants targeting the C-type lectin Mincle. Proc Natl Acad Sci U S A 114, 2675-2680.

Desclozeaux, M., Robbins, A., Jelocnik, M., Khan, S.A., Hanger, J., Gerdts, V., Potter, A., Polkinghorne, A., Timms, P., 2017. Immunization of a wild koala population with a recombinant Chlamydia pecorum Major Outer Membrane Protein (MOMP) or Polymorphic Membrane Protein (PMP) based vaccine: New insights into immune response, protection and clearance. PloS one 12, e0178786.

Dhakal, S., Lu, F., Ghimire, S., Renu, S., Lakshmanappa, Y.S., Hogshead, B.T., Ragland, D., HogenEsch, H., Renukaradhya, G.J., 2019. Corn-derived alpha-D-glucan nanoparticles as adjuvant for intramuscular and intranasal immunization in pigs. Nanomedicine 16, 226-235.

Diaz-San Segundo, F., Dias, C.C., Moraes, M.P., Weiss, M., Perez-Martin, E., Salazar, A.M., Grubman, M.J., de Los Santos, T., 2014. Poly ICLC increases the potency of a replication-defective human adenovirus vectored foot-and-mouth disease vaccine. Virology 468-470, 283-292.

Didierlaurent, A.M., Morel, S., Lockman, L., Giannini, S.L., Bisteau, M., Carlsen, H., Kielland, A., Vosters, O., Vanderheyde, N., Schiavetti, F., Larocque, D., Van Mechelen, M., Garcon, N., 2009. AS04, an aluminum salt- and TLR4 agonist-based adjuvant system, induces a transient localized innate immune response leading to enhanced adaptive immunity. J Immunol 183, 6186-6197.

Dudek, K., Bednarek, D., Ayling, R.D., Kycko, A., Szacawa, E., Karpinska, T.A., 2016. An experimental vaccine composed of two adjuvants gives protection against Mycoplasma bovis in calves. Vaccine 34, 3051-3058.

Duthie, M.S., Windish, H.P., Fox, C.B., Reed, S.G., 2011. Use of defined TLR ligands as adjuvants within human vaccines. Immunol Rev 239, 178-196.

Fang, L., Zhen, Y., Su, Q., Zhu, H., Guo, X., Zhao, P., 2019. Efficacy of CpG-ODN and Freund's immune adjuvants on antibody responses induced by chicken infectious anemia virus VP1, VP2, and VP3 subunit proteins. Poult Sci 98, 1121-1126.

Fossum, C., Hjertner, B., Ahlberg, V., Charerntantanakul, W., McIntosh, K., Fuxler, L., Balagunaseelan, N., Wallgren, P., Lovgren Bengtsson, K., 2014. Early inflammatory response to the saponin adjuvant Matrix-M in the pig. Veterinary immunology and immunopathology 158, 53-61.

Fuchs, L.I., Fort, M.C., Cano, D., Bonetti, C.M., Gimenez, H.D., Vazquez, P.M., Bacigalupe, D., Breccia, J.D., Campero, C.M., Oyhenart, J.A., 2017. Clearance of Tritrichomonas foetus in experimentally infected heifers protected with vaccines based on killed-T. foetus with different adjuvants. Vaccine 35, 1341-1346.

Galliher-Beckley, A., Pappan, L.K., Madera, R., Burakova, Y., Waters, A., Nickles, M., Li, X., Nietfeld, J., Schlup, J.R., Zhong, Q., McVey, S., Dritz, S.S., Shi, J., 2015. Characterization of a novel oil-in-water emulsion adjuvant for swine influenza virus and Mycoplasma hyopneumoniae vaccines. Vaccine 33, 2903-2908.

Gao, X., Xu, K., Yang, G., Shi, C., Huang, H., Wang, J., Yang, W., Liu, J., Liu, Q., Kang, Y., Jiang, Y., Wang, C., 2019. Construction of a novel DNA vaccine candidate targeting F gene of genotype VII Newcastle disease virus and chicken IL-18 delivered by Salmonella. J Appl Microbiol 126, 1362-1372.

Garg, R., Babiuk, L., van Drunen Littel-van den Hurk, S., Gerdts, V., 2017. A novel combination adjuvant platform for human and animal vaccines. Vaccine 35, 4486-4489.

Garg, R., Latimer, L., Gerdts, V., Potter, A., van Drunen Littel-van den Hurk, S., 2015. The respiratory syncytial virus fusion protein formulated with a novel combination adjuvant induces balanced immune responses in lambs with maternal antibodies. Vaccine 33, 1338-1344.

Garg, R., Latimer, L., Simko, E., Gerdts, V., Potter, A., van den Hurk, S., 2014. Induction of mucosal immunity and protection by intranasal immunization with a respiratory syncytial virus subunit vaccine formulation. J Gen Virol 95, 301-306.

Gerdts, V., Mutwiri, G., Richards, J., van Drunen Littel-van den Hurk, S., Potter, A.A., 2013. Carrier molecules for use in veterinary vaccines. Vaccine 31, 596-602.

Glenny, A.T.P., C.G.; Waddington, H., and Wallace, V., 1926. The antigenic value of toxoid precipitated by potassium-alum. J pathol bacteriol 29, 38-45.

Gregory, A.E., Titball, R., Williamson, D., 2013. Vaccine delivery using nanoparticles. Frontiers in cellular and infection microbiology 3, 13.

Gupta, S.K., Bajwa, P., Deb, R., Chellappa, M.M., Dey, S., 2014. Flagellin a toll-like receptor 5 agonist as an adjuvant in chicken vaccines. Clin Vaccine Immunol 21, 261-270.

Hajam, I.A., Dar, P.A., Appavoo, E., Kishore, S., Bhanuprakash, V., Ganesh, K., 2015. Bacterial Ghosts of Escherichia coli Drive Efficient Maturation of Bovine Monocyte-Derived Dendritic Cells. PloS one 10, e0144397.

Hajam, I.A., Dar, P.A., Shahnawaz, I., Jaume, J.C., Lee, J.H., 2017. Bacterial flagellin-a potent immunomodulatory agent. Exp Mol Med 49, e373.

Harte, C., Gorman, A.L., McCluskey, S., Carty, M., Bowie, A.G., Scott, C.J., Meade, K.G., Lavelle, E.C., 2017. Alum Activates the Bovine NLRP3 Inflammasome. Frontiers in immunology 8, 1494.

Hiremath, J., Kang, K.I., Xia, M., Elaish, M., Binjawadagi, B., Ouyang, K., Dhakal, S., Arcos, J., Torrelles, J.B., Jiang, X., Lee, C.W., Renukaradhya, G.J., 2016. Entrapment of H1N1 Influenza Virus Derived Conserved Peptides in PLGA Nanoparticles Enhances T Cell Response and Vaccine Efficacy in Pigs. PloS one 11, e0151922.

HogenEsch, H., O'Hagan, D.T., Fox, C.B., 2018. Optimizing the utilization of aluminum adjuvants in vaccines: you might just get what you want. NPJ vaccines 3, 51.

Ishikawa, E., Ishikawa, T., Morita, Y.S., Toyonaga, K., Yamada, H., Takeuchi, O., Kinoshita, T., Akira, S., Yoshikai, Y., Yamasaki, S., 2009. Direct recognition of the mycobacterial glycolipid, trehalose dimycolate, by C-type lectin Mincle. J Exp Med 206, 2879-2888.

Jones, G.J., Steinbach, S., Clifford, D., Baldwin, S.L., Ireton, G.C., Coler, R.N., Reed, S.G., Vordermeier, H.M., 2013. Immunisation with ID83 fusion protein induces antigen-specific cell mediated and humoral immune responses in cattle. Vaccine 31, 5250-5255.

Kass, P.H., Barnes, W.G., Jr., Spangler, W.L., Chomel, B.B., Culbertson, M.R., 1993. Epidemiologic evidence for a causal relation between vaccination and fibrosarcoma tumorigenesis in cats. J Am Vet Med Assoc 203, 396-405.

Kensil, C.R., 1996. Saponins as vaccine adjuvants. Crit Rev Ther Drug Carrier Syst 13, 1-55.

Khan, S.A., Waugh, C., Rawlinson, G., Brumm, J., Nilsson, K., Gerdts, V., Potter, A., Polkinghorne, A., Beagley, K., Timms, P., 2014. Vaccination of koalas (Phascolarctos cinereus) with a recombinant chlamydial major outer membrane protein adjuvanted with poly I:C, a host defense peptide and polyphosphazine, elicits strong and long lasting cellular and humoral immune responses. Vaccine 32, 5781-5786.

Kim, S., Kaiser, P., Borowska, D., Vervelde, L., 2018. Synergistic effect of co-stimulation of membrane and endosomal TLRs on chicken innate immune responses. Veterinary immunology and immunopathology 199, 15-21.

Kool, M., Fierens, K., Lambrecht, B.N., 2012. Alum adjuvant: some of the tricks of the oldest adjuvant. Journal of medical microbiology 61, 927-934.

Korsholm, K.S., Andersen, P.L., Christensen, D., 2012. Cationic liposomal vaccine adjuvants in animal challenge models: overview and current clinical status. Expert review of vaccines 11, 561-577.

Kovacs-Nolan, J., Latimer, L., Landi, A., Jenssen, H., Hancock, R.E., Babiuk, L.A., van Drunen Littel-van den Hurk, S., 2009a. The novel adjuvant combination of CpG ODN, indolicidin and polyphosphazene induces potent antibody- and cell-mediated immune responses in mice. Vaccine 27, 2055-2064.

Kovacs-Nolan, J., Mapletoft, J.W., Lawman, Z., Babiuk, L.A., van Drunen Littel-van den Hurk, S., 2009b. Formulation of bovine respiratory syncytial virus fusion protein with CpG oligodeoxynucleotide, cationic host defence peptide and polyphosphazene enhances humoral and cellular responses and induces a protective type 1 immune response in mice. J Gen Virol 90, 1892-1905.

Lacasta, A., Mwalimu, S., Kibwana, E., Saya, R., Awino, E., Njoroge, T., Poole, J., Ndiwa, N., Pelle, R., Nene, V., Steinaa, L., 2018. Immune parameters to p67C antigen adjuvanted with ISA206VG correlate with protection against East Coast fever. Vaccine 36, 1389-1397.

Levast, B., Awate, S., Babiuk, L., Mutwiri, G., Gerdts, V., van Drunen Littel-van den Hurk, S., 2014. Vaccine Potentiation by Combination Adjuvants. Vaccines 2, 297-322.

Li, Z., Tang, X., Suo, J., Qin, M., Yin, G., Liu, X., Suo, X., 2015. Transgenic Eimeria mitis expressing chicken interleukin 2 stimulated higher cellular immune response in chickens compared with the wild-type parasites. Front Microbiol 6, 533.

Ma, X., Bi, S., Wang, Y., Chi, X., Hu, S., 2019. Combined adjuvant effect of ginseng stem-leaf saponins and selenium on immune responses to a live bivalent vaccine of Newcastle disease virus and infectious bronchitis virus in chickens. Poult Sci 98, 3548-3556.

Magiri, R., Lai, K., Huang, Y., Mutwiri, G., Wilson, H.L., 2019. Innate immune response profiles in pigs injected with vaccine adjuvants polydi(sodium carboxylatoethylphenoxy)phosphazene (PCEP) and Emulsigen. Vet Immunol Immunopathol 209, 7-16.

Magiri, R.B., Lai, K., Chaffey, A.M., Wilson, H.L., Berry, W.E., Szafron, M.L., Mutwiri, G.K., 2016. Response of immune response genes to adjuvants poly [di(sodium carboxylatoethylphenoxy)phosphazene] (PCEP), CpG oligodeoxynucleotide and emulsigen at intradermal injection site in pigs. Vet Immunol Immunopathol 175, 57-63.

Mahony, D., Mody, K.T., Cavallaro, A.S., Hu, Q., Mahony, T.J., Qiao, S., Mitter, N., 2015. Immunisation of Sheep with Bovine Viral Diarrhoea Virus, E2 Protein Using a Freeze-Dried Hollow Silica Mesoporous Nanoparticle Formulation. PloS one 10, e0141870.

Mair, K.H., Koinig, H., Gerner, W., Hohne, A., Bretthauer, J., Kroll, J.J., Roof, M.B., Saalmuller, A., Stadler, K., Libanova, R., 2015. Carbopol improves the early cellular immune responses induced by the modified-life vaccine Ingelvac PRRS(R) MLV. Vet Microbiol 176, 352-357.

Maisonneuve, C., Bertholet, S., Philpott, D.J., De Gregorio, E., 2014. Unleashing the potential of NOD- and Toll-like agonists as vaccine adjuvants. Proceedings of the National Academy of Sciences of the United States of America 111, 12294-12299.

Makadiya, N., Brownlie, R., van den Hurk, J., Berube, N., Allan, B., Gerdts, V., Zakhartchouk, A., 2016. S1 domain of the porcine epidemic diarrhea virus spike protein as a vaccine antigen. Virol J 13, 57.

Marrack, P., McKee, A.S., Munks, M.W., 2009. Towards an understanding of the adjuvant action of aluminium. Nature reviews. Immunology 9, 287-293.

Matoo, J.J., Bashir, K., Kumar, A., Krishnaswamy, N., Dey, S., Chellappa, M.M., Ramakrishnan, S., 2018. Resiquimod enhances mucosal and systemic immunity against avian infectious bronchitis virus vaccine in the chicken. Microb Pathog 119, 119-124.

McKay, P.F., King, D.F., Mann, J.F., Barinaga, G., Carter, D., Shattock, R.J., 2016. TLR4 and TLR7/8 Adjuvant Combinations Generate Different Vaccine Antigen-Specific Immune Outcomes in Minipigs when Administered via the ID or IN Routes. PLoS One 11, e0148984.

Mody, K.T., Mahony, D., Cavallaro, A.S., Zhang, J., Zhang, B., Mahony, T.J., Yu, C., Mitter, N., 2015. Silica Vesicle Nanovaccine Formulations Stimulate Long-Term Immune Responses to the Bovine Viral Diarrhoea Virus E2 Protein. PloS one 10, e0143507.

Mody, K.T., Mahony, D., Zhang, J., Cavallaro, A.S., Zhang, B., Popat, A., Mahony, T.J., Yu, C., Mitter, N., 2014. Silica vesicles as nanocarriers and adjuvants for generating both antibody and T-cell mediated immune resposes to Bovine Viral Diarrhoea Virus E2 protein. Biomaterials 35, 9972-9983.

Morel, S., Didierlaurent, A., Bourguignon, P., Delhaye, S., Baras, B., Jacob, V., Planty, C., Elouahabi, A., Harvengt, P., Carlsen, H., Kielland, A., Chomez, P., Garcon, N., Van Mechelen, M., 2011. Adjuvant System AS03 containing alpha-tocopherol modulates innate immune response and leads to improved adaptive immunity. Vaccine 29, 2461-2473.

Mosca, F., Tritto, E., Muzzi, A., Monaci, E., Bagnoli, F., Iavarone, C., O'Hagan, D., Rappuoli, R., De Gregorio, E., 2008. Molecular and cellular signatures of human vaccine adjuvants. Proceedings of the National Academy of Sciences of the United States of America 105, 10501-10506.

Neeland, M.R., Elhay, M.J., Meeusen, E.N., de Veer, M.J., 2014a. Vaccination with liposomal poly(I:C) induces discordant maturation of migratory dendritic cell subsets and anti-viral gene signatures in afferent lymph cells. Vaccine 32, 6183-6192.

Neeland, M.R., Elhay, M.J., Nathanielsz, J., Meeusen, E.N., de Veer, M.J., 2014b. Incorporation of CpG into a liposomal vaccine formulation increases the maturation of antigen-loaded dendritic cells and monocytes to improve local and systemic immunity. J Immunol 192, 3666-3675.

Nichani, A.K., Kaushik, R.S., Mena, A., Popowych, Y., Dent, D., Townsend, H.G., Mutwiri, G., Hecker, R., Babiuk, L.A., Griebel, P.J., 2004. CpG oligodeoxynucleotide induction of antiviral effector molecules in sheep. Cell Immunol 227, 24-37.

Oleszycka, E., Lavelle, E.C., 2014. Immunomodulatory properties of the vaccine adjuvant alum. Current opinion in immunology 28, 1-5.

Park, C., Jeong, J., Choi, K., Park, S.J., Kang, I., Chae, C., 2017. Development of porcine circovirus 2 (PCV2) open reading frame 2 DNA vaccine with different adjuvants and comparison with commercial PCV2 subunit vaccine in an experimental challenge. Canadian journal of veterinary research = Revue canadienne de recherche veterinaire 81, 171-177.

Park, M.E., Lee, S.Y., Kim, R.H., Ko, M.K., Lee, K.N., Kim, S.M., Kim, B.K., Lee, J.S., Kim, B., Park, J.H., 2014. Enhanced immune responses of foot-and-mouth disease vaccine using new oil/gel adjuvant mixtures in pigs and goats. Vaccine 32, 5221-5227.

Quattrocchi, V., Pappalardo, J.S., Langellotti, C., Smitsaart, E., Fondevila, N., Zamorano, P., 2014. Early protection against foot-and-mouth disease virus in cattle using an inactivated vaccine formulated with Montanide ESSAI IMS D 12802 VG PR adjuvant. Vaccine 32, 2167-2172.

Quattrocchi, V., Soria, I., Langellotti, C.A., Gnazzo, V., Gammella, M., Moore, D.P., Zamorano, P.I., 2017. A DNA Vaccine Formulated with Chemical Adjuvant Provides Partial Protection against Bovine Herpes Virus Infection in Cattle. Frontiers in immunology 8, 37.

Querec, T., Bennouna, S., Alkan, S., Laouar, Y., Gorden, K., Flavell, R., Akira, S., Ahmed, R., Pulendran, B., 2006. Yellow fever vaccine YF-17D activates multiple dendritic cell subsets via TLR2, 7, 8, and 9 to stimulate polyvalent immunity. The Journal of experimental medicine 203, 413-424.

Rajput, Z.I., Hu, S.H., Xiao, C.W., Arijo, A.G., 2007. Adjuvant effects of saponins on animal immune responses. J Zhejiang Univ Sci B 8, 153-161.

Ramon, G., 1924. Sur la toxine et sur l'anatoxine diphtheriques. Ann Inst Pasteur 38, 1-10.

Rankin, R., Pontarollo, R., Gomis, S., Karvonen, B., Willson, P., Loehr, B.I., Godson, D.L., Babiuk, L.A., Hecker, R., van Drunen Littel-van den Hurk, S., 2002. CpG-containing oligodeoxynucleotides augment and switch the immune responses of cattle to bovine herpesvirus-1 glycoprotein D. Vaccine 20, 3014-3022.

Reed, S.G., Bertholet, S., Coler, R.N., Friede, M., 2009. New horizons in adjuvants for vaccine development. Trends Immunol 30, 23-32.

Riitho, V., Walters, A.A., Somavarapu, S., Lamp, B., Rumenapf, T., Krey, T., Rey, F.A., Oviedo-Orta, E., Stewart, G.R., Locker, N., Steinbach, F., Graham, S.P., 2017. Design and evaluation of the immunogenicity and efficacy of a biomimetic particulate formulation of viral antigens. Scientific reports 7, 13743.

Sachan, S., Ramakrishnan, S., Annamalai, A., Sharma, B.K., Malik, H., Saravanan, B.C., Jain, L., Saxena, M., Kumar, A., Krishnaswamy, N., 2015. Adjuvant potential of resiquimod with inactivated Newcastle disease vaccine and its mechanism of action in chicken. Vaccine 33, 4526-4532.

Sadat, S.M.A., Snider, M., Garg, R., Brownlie, R., van Drunen Littel-van den Hurk, S., 2017. Local innate responses and protective immunity after intradermal immunization with bovine viral diarrhea virus E2 protein formulated with a combination adjuvant in cattle. Vaccine 35, 3466-3473.

Sanders, M.T., Brown, L.E., Deliyannis, G., Pearse, M.J., 2005. ISCOM-based vaccines: the second decade. Immunology and cell biology 83, 119-128.

Sarkar, I., Garg, R., van Drunen Littel-van den Hurk, S., 2016. Formulation of the respiratory syncytial virus fusion protein with a polymer-based combination adjuvant promotes transient and local innate immune responses and leads to improved adaptive immunity. Vaccine 34, 5114-5124.

Sarkar, I., Garg, R., van Drunen Littel-van den Hurk, S., 2018. The respiratory syncytial virus fusion protein formulated with a polymer-based adjuvant induces multiple signaling pathways in macrophages. Vaccine 36, 2326-2336.

Schmidt, S.T., Pedersen, G.K., Neustrup, M.A., Korsholm, K.S., Rades, T., Andersen, P., Foged, C., Christensen, D., 2018. Induction of Cytotoxic T-Lymphocyte Responses Upon Subcutaneous Administration of a Subunit Vaccine Adjuvanted With an Emulsion Containing the Toll-Like Receptor 3 Ligand Poly(I:C). Frontiers in immunology 9, 898.

Schoenen, H., Bodendorfer, B., Hitchens, K., Manzanero, S., Werninghaus, K., Nimmerjahn, F., Agger, E.M., Stenger, S., Andersen, P., Ruland, J., Brown, G.D., Wells, C., Lang, R., 2010. Cutting edge: Mincle is essential for recognition and adjuvanticity of the mycobacterial cord factor and its synthetic analog trehalose-dibehenate. J Immunol 184, 2756-2760.

Shah, R.R., O'Hagan, D.T., Amiji, M.M., Brito, L.A., 2014. The impact of size on particulate vaccine adjuvants. Nanomedicine (Lond) 9, 2671-2681.

Singh, M., O'Hagan, D.T., 2003. Recent advances in veterinary vaccine adjuvants. Int J Parasitol 33, 469-478.

Snider, M., Garg, R., Brownlie, R., van den Hurk, J.V., van Drunen Littel-van den Hurk, S., 2014. The bovine viral diarrhea virus E2 protein formulated with a novel adjuvant induces strong, balanced immune responses and provides protection from viral challenge in cattle. Vaccine 32, 6758-6764.

Song, L., Xiong, D., Hu, M., Jiao, X., Pan, Z., 2018. Enhanced Th1/Th2 mixed immune responses elicited by polyethyleneimine adjuvanted influenza A (H7N9) antigen HA1-2 in chickens. Poult Sci 97, 4245-4251.

Spickler, A.R., Roth, J.A., 2003. Adjuvants in veterinary vaccines: modes of action and adverse effects. J Vet Intern Med 17, 273-281.

Stills, H.F., Jr., 2005. Adjuvants and antibody production: dispelling the myths associated with Freund's complete and other adjuvants. ILAR J 46, 280-293.

Sun, B., Yu, S., Zhao, D., Guo, S., Wang, X., Zhao, K., 2018. Polysaccharides as vaccine adjuvants. Vaccine 36, 5226-5234.

Sun, H.X., Xie, Y., Ye, Y.P., 2009. Advances in saponin-based adjuvants. Vaccine 27, 1787-1796.

Tabynov, K., Sansyzbay, A., Tulemissova, Z., Tabynov, K., Dhakal, S., Samoltyrova, A., Renukaradhya, G.J., Mambetaliyev, M., 2016. Inactivated porcine reproductive and respiratory syndrome virus vaccine adjuvanted with Montanide Gel 01 ST elicits virus-specific cross-protective inter-genotypic response in piglets. Vet Microbiol 192, 81-89.

Tewari, K., Flynn, B.J., Boscardin, S.B., Kastenmueller, K., Salazar, A.M., Anderson, C.A., Soundarapandian, V., Ahumada, A., Keler, T., Hoffman, S.L., Nussenzweig, M.C., Steinman, R.M., Seder, R.A., 2010. Poly(I:C) is an effective adjuvant for antibody and multi-functional CD4+ T cell responses to Plasmodium falciparum circumsporozoite protein (CSP) and alphaDEC-CSP in non human primates. Vaccine 28, 7256-7266.

Thakur, A., Andrea, A., Mikkelsen, H., Woodworth, J.S., Andersen, P., Jungersen, G., Aagaard, C., 2018. Targeting the Mincle and TLR3 receptor using the dual agonist cationic adjuvant formulation 9 (CAF09) induces humoral and polyfunctional memory T cell responses in calves. PloS one 13, e0201253.

Toussi, D.N., Massari, P., 2014. Immune Adjuvant Effect of Molecularly-defined Toll-Like Receptor Ligands. Vaccines 2, 323-353.

Ugolini, M., Gerhard, J., Burkert, S., Jensen, K.J., Georg, P., Ebner, F., Volkers, S.M., Thada, S., Dietert, K., Bauer, L., Schafer, A., Helbig, E.T., Opitz, B., Kurth, F., Sur, S., Dittrich, N., Gaddam, S., Conrad, M.L., Benn, C.S., Blohm, U., Gruber, A.D., Hutloff, A., Hartmann, S., Boekschoten, M.V., Muller, M., Jungersen, G., Schumann, R.R., Suttorp, N., Sander, L.E., 2018. Recognition of microbial viability via TLR8 drives TFH cell differentiation and vaccine responses. Nature immunology 19, 386-396.

Van Donkersgoed, J., McCartney, D., S., v.D.L.v.d.H., 1996. Efficacy of an experimental BHV 1 subunit gIV vaccine in beef calves challenged with BHV 1 in aerosol. Can. J. Vet. Res. 60, 55-58.

Vreman, S., Auray, G., Savelkoul, H.F.J., Rebel, A., Summerfield, A., Stockhofe-Zurwieden, N., 2018. Neonatal porcine blood derived dendritic cell subsets show activation after TLR2 or TLR9 stimulation. Developmental and comparative immunology 84, 361-370.

Wang, Y.Q., Liu, Y., Wang, Y.X., Wu, Y.J., Jia, P.Y., Shan, J.J., Wu, J., Ma, G.H., Su, Z.G., 2016. The potential adjuvanticity of quaternized chitosan hydrogel based microparticles for porcine reproductive and respiratory syndrome virus inactivated vaccine. Int Immunopharmacol 39, 84-91.

Werninghaus, K., Babiak, A., Gross, O., Holscher, C., Dietrich, H., Agger, E.M., Mages, J., Mocsai, A., Schoenen, H., Finger, K., Nimmerjahn, F., Brown, G.D., Kirschning, C., Heit, A., Andersen, P., Wagner, H., Ruland, J., Lang, R., 2009. Adjuvanticity of a synthetic cord factor analogue for subunit Mycobacterium tuberculosis vaccination requires FcRgamma-Syk-Card9-dependent innate immune activation. J Exp Med 206, 89-97.

Wu, H.W., Fu, Z.Q., Lu, K., Pond-Tor, S., Meng, R., Hong, Y., Chu, K., Li, H., Jiz, M., Liu, J.M., Hou, M., Park, S., Lin, J.J., Kurtis, J.D., 2017. Vaccination with recombinant paramyosin in Montanide ISA206 protects against Schistosoma japonicum infection in water buffalo. Vaccine 35, 3409-3415.

Xiong, Q., Wei, Y., Xie, H., Feng, Z., Gan, Y., Wang, C., Liu, M., Bai, F., Xie, F., Shao, G., 2014. Effect of different adjuvant formulations on the immunogenicity and protective effect of a live Mycoplasma hyopneumoniae vaccine after intramuscular inoculation. Vaccine 32, 3445-3451.

Yeung, A., Gellatly, S., Hancock, R., 2011. Multifunctional cationic host defence peptides and their clinical applications. Cellular and Molecular Life Sciences, 1-16.

Zhang, B., Cavallaro, A.S., Mody, K.T., Zhang, J., Deringer, J.R., Brown, W.C., Mahony, T.J., Yu, C., Mitter, N., 2016a. Nanoparticle-Based Delivery of Anaplasma marginale Membrane Proteins; VirB9-1 and VirB10 Produced in the Pichia pastoris Expression System. Nanomaterials (Basel) 6.

Zhang, T., Liu, Y., Wang, H., Zhang, X., Zhu, S., Xu, P., Yin, J., Ren, G., Liu, J., Li, D., 2016b. Recombinant NDV expressing cytokines or fliC confers a quick immune response against NDV challenge and resistance to maternal antibody. Vet Microbiol 196, 107-117.

Zhang, Y., Palmer, G.H., Abbott, J.R., Howard, C.J., Hope, J.C., Brown, W.C., 2003. CpG ODN 2006 and IL-12 are comparable for priming Th1 lymphocyte and IgG responses in cattle immunized with a rickettsial outer membrane protein in alum. Vaccine 21, 3307-3318.

Zhang, Y., Wang, L., Ruan, W., Zhang, J., Yao, P., Zhou, S., An, J., 2014. Immunization with recombinant 3-1E protein in AbISCO(R)-300 adjuvant induced protective immunity against Eimeria acervulina infection in chickens. Exp Parasitol 141, 75-81.

Zhao, K., Li, S., Li, W., Yu, L., Duan, X., Han, J., Wang, X., Jin, Z., 2017. Quaternized chitosan nanoparticles loaded with the combined attenuated live vaccine against Newcastle disease and infectious bronchitis elicit immune response in chicken after intranasal administration. Drug Deliv 24, 1574-1586.

Zhu, F., Liu, X., Sun, Z., Yu, C., Liu, L., Yang, S., Li, B., Wei, K., Zhu, R., 2016. Immune-Enhancing Effects of Taishan Pinus massoniana Pollen Polysaccharides on DNA Vaccine Expressing Bordetella avium ompA. Front Microbiol 7, 66.

Chapter 5

Challenges in Having Vaccines Available to Control Transboundary Diseases of Livestock

Charles E. Lewis[1,2] and James A. Roth[1,2*]

[1]Department of Veterinary Microbiology and Preventive Medicine, College of Veterinary Medicine, Iowa State University, Ames, Iowa, USA
[2]Interdepartmental Microbiology Program, College of Veterinary Medicine, Iowa State University, Ames, Iowa, USA

*Corresponding authors: jaroth@iastate.edu

DOI: https://doi.org/10.21775/9781913652593.05

Abstract

The global human population is growing at a rapid rate leading to the need for continued expansion of food animal production to meet the world's increasing nutritional requirements. As a consequence of this increased production demand, the use of high volume, animal dense systems have expanded providing high quality protein at reduced costs. Backyard animal production has also expanded. This increased food animal production has facilitated the rapid spread, mutation, and adaptation of pathogens to new hosts. This scenario continues to drive the emergence and reemergence of diseases in livestock species increasing the urgency for development and availability of vaccines for transboundary animal diseases (TADs). Even though vaccines are widely recognized as being an essential tool for control of TADs, there are many scientific, economic, political, and logistical challenges to having vaccine available to control an outbreak. This chapter will focus on examples of the challenges associated with having vaccines available for emergency response, as well as the characteristics of "ideal" TAD vaccines, the need for complementary diagnostic assays, and hurdles involved in bringing efficacious veterinary TAD vaccines to market including regulatory constraints and considerations for stockpiling vaccines for emergency use in non-endemic countries.

Examples will also highlight the complicated interplay between animal health and human health and demonstrate the lasting benefits that can be gained from an efficacious vaccine.

Introduction

As the global human population is growing at an exponential rate and expected to reach 9.8 billion people in 2050, there has been a corresponding, unprecedented expansion and growth of food animal production to meet the increasing protein demand to support human nutrition requirements (Roth, 2019). The United Nations Food and Agriculture Organization (FAO) estimates that, to meet this increased demand by 2050, food production must increase by approximately 70% compared to 2005 production levels (Figure 1) (FAO., 2009). As a consequence of the expansion of both the global human and production animal populations, there has been an increase in the emergence and reemergence of animal diseases (Hassell et al., 2017). This increased demand has driven the trend towards high volume, animal dense systems which provide ample opportunity for the rapid spread of disease, as well as increased fitness pressure leading to mutation and adaptation of pathogens to new host species or increased disease severity in historically susceptible populations (Hassell et al., 2017). Livestock and poultry are a significant contributors to the economy of many countries, accounting for over half of United States agricultural cash receipts, often exceeding $100 billion per year. The increasing human population is also leading to an increase in "backyard" animal production in both rural and urban environments. This can be an important source of high-quality protein and income for food insecure households (Roth, 2011). However, the backyard animals typically do not benefit from biosecurity, vaccines or other veterinary care and can be a source of infection with zoonotic diseases. The increase in concentrated animal feeding operations and in backyard animal production to feed the increasing human population increases the urgency for development and availability of vaccines for transboundary animal diseases.

Transboundary Animal Diseases (TADs) are a major threat to agriculture animal health, as well as human health, and are defined by the United Nations Food and Agriculture Organization (FAO) as those diseases that have the potential meet a combination of the following criteria: (1) they are highly contagious; (2) they have the potential to spread rapidly, irrespective of national borders; (3) they are associated

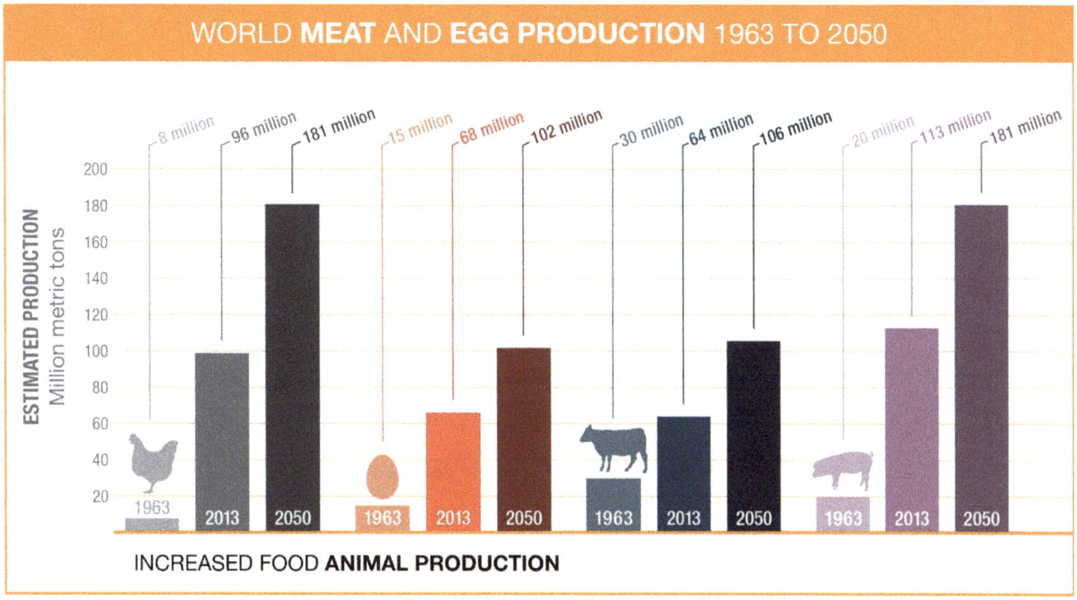

Figure 1. The United Nations Food and Agriculture Organization estimated world meat and egg production will need to increase by 70% from 2005/2007 to 2050 to meet the needs of the growing population. From Roth, 2019.

with high mortality and/or morbidity in affected species; (4) they have the potential to affect food security and cause severe socioeconomic consequences; (5) they may be associated with public health concerns (FAO, 1996; Rossiter and Al Hammadi, 2009). Historically, TADs have had devastating consequences on both human and animal populations. Details of a few of these accounts will be discussed later in this chapter.

With increased globalization, the introduction of TADs into non-endemic areas can occur in a variety of ways. One of the most important translocation mechanisms include the domestic and global movement of diseased animals and contaminated animal products either as imports or food waste (Otte et al., 2004). Other means of introduction can occur through the importation of contaminated biological products or germplasm, the movement of people, and the migration of affected animal species. Climatic changes can lead to the natural spread of insect vectors and, in the case of some diseases, shifts in wind currents can lead to the spread of diseases (Jones et al., 2013). A TAD can rapidly become endemic in introduced areas threatening naive contact countries, which increases the likelihood of continued spread and further difficulty in gaining control of the outbreak, ultimately leading to greater impacts on food security (Otte et al., 2004). This dire situation is exacerbated in situations in which an efficacious and cost-effective vaccine is not available. A prime example of how rapidly and widely a disease can disperse without an available vaccine to implement as a control measure can be seen in the spread of African swine fever (ASF) disease in eastern Europe and Asia. ASF will be discussed in detail later in this chapter, as well as discussion on Foot-and-Mouth Disease (FMD), as these two diseases will be compared in regards to strategies towards developing "ideal" TAD vaccine platforms and their potential use as a component to control the impact of the disease.

There are numerous tools and approaches utilized to control the spread and damage created by the incursion of a TAD into a non-endemic area. These can include the control of movement of animals and people between areas, implementing and heightening biosecurity at the farm level, depopulation of infected and in-contact herds, epidemiologic trace back and trace forward of cases, implementation of rapid diagnostics, and mass vaccination. Implementation of vaccination campaigns are based on the overall objective and can include vaccinate-to-kill, vaccinate-to-slaughter, or vaccinate-to-live strategies (Barnett et al., 2015b; Grubman and Baxt, 2004). The availability of TAD vaccines is of paramount importance for maintaining and advancing animal health, public health, and food production and security. They serve as a cost-effective methods to control and prevent animal disease, lead to increased high-quality protein availability, expand the efficiency of food production, as well as functioning to reduce or prevent the transmission of zoonotic and foodborne infections to people (Roth, 2011).

Even though vaccines are widely recognized as being an essential tool for control of TADs, there are many scientific, economic, political, and logistical challenges to having vaccine available to control a TAD outbreak. In this chapter we will discuss three transboundary animal diseases as different examples of the challenges associated with having vaccines available for emergency response to TADs (foot and mouth disease, African swine fever, and Nipah virus infection). The chapter will also focus on the characteristics of "ideal" TAD vaccines and the need for complementary diagnostic assays, hurdles involved in bringing efficacious veterinary TAD vaccines to market including regulatory constraints, and considerations for stockpiling vaccines for emergency use in non-endemic countries. The chapter will conclude with the example of the use of a veterinary vaccine for the successful, global eradication of rinderpest and its relation to increased food security and indirect impacts on human disease and public health. This will highlight the complicated interplay between animal health and human health and demonstrate the lasting benefits that can be gained from an efficacious vaccine. The rinderpest example will also demonstrate how introduction of a TAD can lead to famine, increased incidence of indirectly related disease in humans, and ultimately lead to the death of people and millions of animals.

Foot-and-Mouth Disease

As a transboundary animal disease, Foot and Mouth Disease (FMD) is perhaps the most important pathogen of livestock worldwide. It is a highly contagious, clinically acute, viral disease affecting members of *Artiodactyla* (cloven-hoofed animals), caused by the *Foot and Mouth Disease Virus* (FMDV), a member of the *Apthovirus* genus in the family Picornaviridae (Rweyemamu et al., 2008). The effects of this disease are of greatest concern, at least to the production livestock industry, in cattle and domestic swine, though disease has been reported in at least 70 species of wild animals, including wild ruminants and feral swine, which can play a significant role in its epidemiology (Alexandrov et al., 2013; Knight-Jones et al., 2016; Lee et al., 2017; Thompson et al., 2002; Thomson et al., 2003).

FMDV is a complicated virus with seven serotypes all having varying geographic distribution (O, A, C, Asia 1, SAT 1, SAT 2, and SAT 3) (Figure 2) (Grubman and Baxt, 2004; Rweyemamu et al., 2008). Globally, serotype O is the most common and was responsible for a pan-Asian outbreak that began in 1990 and has subsequently engulfed many countries throughout the world. Serotype C is considered to be eradicated, but is still utilized in the formulation of some multivalent vaccines (Jamal and Belsham, 2013). There are inconsistent levels of variability between the FMD serotypes and they collectively contain more than 60 viral strains with novel strains arising regularly. Continued circulation of the virus in the field and the quasispecies nature of the virus genome leads to the continued rise of new strains, which can continue to evolve to break-through vaccine-induced immunity (Grubman and Baxt,

2004; Haydon et al., 2001; Jamal and Belsham, 2013). The vast majority of strains affect all susceptible species, yet some have more restricted host ranges (Arzt et al., 2011a). For instance, the O Cathay strain appears to only affect pigs. Immunity to one serotype does not reliably convey protection from other serotypes and intra-serotype strain protection varies with the level of antigenic similarity (Lee et al., 2017; Rodriguez and Grubman, 2009).

The FMD virus can be found in most secretions and excretions from acutely affected animals, including expired air with animal species and virus strain accounting for differences in the amount of virus shed by each route. Animals are typically infected by inhalation, ingestion, or through abrasions in the skin or mucous membranes (Bartley et al., 2002; Grubman and Baxt, 2004). Pigs tend to produce large amounts of aerosolized virus, and the presence of a large, high-density swine herd may increase the risk of airborne spread to other animals. As pigs require much higher doses to be infected by inhalation, they are typically infected via ingestion of virus-contaminated material, while cattle are particularly susceptible to infection by the inhalation route (Arzt et al., 2011b; Stenfeldt et al., 2016). Mechanical transmission via fomites plays an important role in spread of this virus and people can also act as mechanical vectors by transporting the virus on their skin or clothing (Amass et al., 2004).

FMD is characterized by high morbidity and low mortality, though it does occasionally cause a fatal acute myocarditis that is typically limited to young animals. The hallmark clinical signs in affected species include a rapidly spreading acute onset, febrile illness with the formation of vesicular lesions on the tongue, dental pad, coronary band and teats (Arzt et al., 2011a; Stenfeldt et al., 2016). Though animals can and do recover from this disease, secondary bacterial infections of the vesicles can lead to chronic lameness, wasting, and mortality (Grubman and Baxt, 2004). As it affects production species, the presence of this disease in a domestic herd can lead to drastically lower yields and have devastating economic consequences. Resulting chronic lameness and permanent hoof damage can have direct impacts on the use of draft animals in many parts of the world where crop yields are dependent on animal labor (Knight-Jones and Rushton, 2013). Although adult animals typically recover, the associated morbidity is significant in naïve populations, including the presence of significant pain and distress in some animal species. Sequelae may

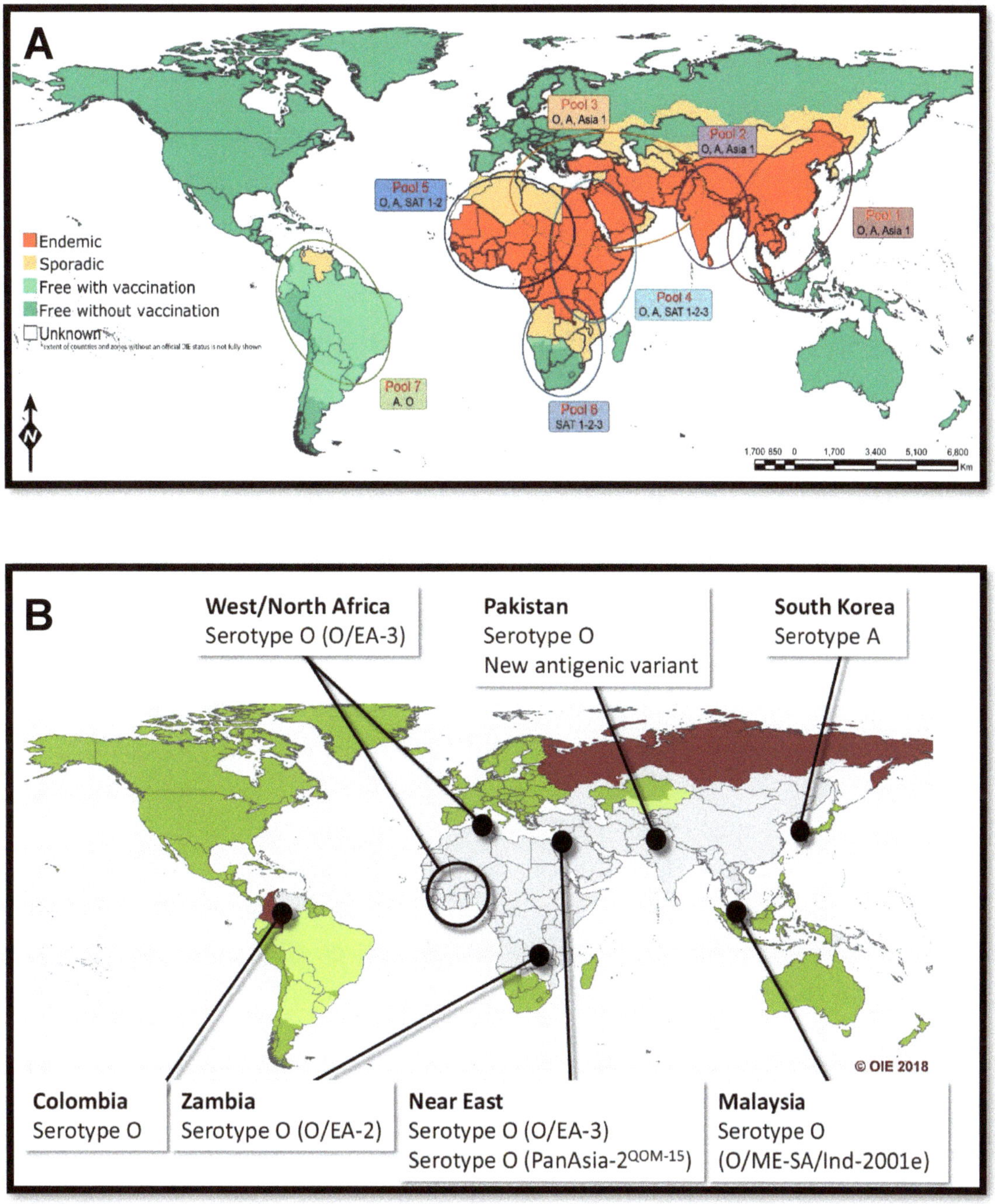

Figure 2. Geographic distribution of the seven serotypes of foot-and-mouth disease virus with the proposed disease status of individual countries in 2018 (A) and the major FMD events from that year (B). From King et al., 2018.

include decreased milk yield, permanent hoof damage, heat-intolerance syndrome, and chronic mastitis (Grubman and Baxt, 2004).

After once being found nearly worldwide, FMD has now been eradicated from 70 countries (including North America, most of South America, Australia, Greenland, Iceland, and western Europe) through the control of movement of susceptible species and animal by-product movement and through the use of intricate mass vaccination and culling campaigns (Brito et al., 2017; Jamal and Belsham, 2013). FMD has not been reported in North America for more than 90 years, with the last outbreak occurring in 1929 (Bachrach, 1968). Even with the success of elimination, there are still over a hundred countries considered endemically or sporadically infected with the disease. Where FMD is endemic, animal production and performance are affected and these countries suffer major economic constraints secondary to trade embargoes that drastically reduced access to international markets. The potential effects of disease on smallholder farms can be significant. Nampanya et al estimated that losses associated with FMD could approach 60% of annual farm-based household income in Laos (Nampanya et al., 2016). Across all FMD-endemic countries, it is estimated that the disease costs \$6.5-\$21billion US dollars (USD) annually in production losses (Knight-Jones and Rushton, 2013).

Outbreaks of FMD in countries that have previously eliminated the disease can be economically devastating due to the loss of trade and the cost of eradication with direct and indirect losses reaching into the billions of US dollars. The 2001 outbreak in the United Kingdom was estimated to cost \$12 billion USD and more than four million cattle and sheep were destroyed in order to reestablish disease free status and regain access to international markets (Thompson et al., 2002). A risk assessment estimated the hypothetical economic consequences of FMDV in the US to be up to \$140 billion USD with most of the cost impacting agricultural firms and consumers (Pendell et al., 2015).

African Swine Fever

Similar to FMD, African swine fever (ASF) is another high-consequence, transboundary animal disease causing great concern throughout the world swine industry. *African Swine Fever Virus* (ASFV) is a tick-borne *Asfivirus*, the sole member of the *Asfarviridae* family, and is the only known DNA arbovirus. The virus is large and complicated, containing over 170 proteins (Figure 3) (Chapman et al., 2008; Dixon et al., 2013; Dixon et al., 2005). ASFV is currently causing an unprecedented outbreak after its introduction and spread from the country of Georgia in both domestic and wild pigs, through Western Europe, and into Russia, ultimately leading to introduction into China, affecting the world's largest swine population and leading to the death of millions of domestic pigs (Dixon et al., 2005). It has been predicted this outbreak has resulted in a deficit of 25 million metric tons of pork in China in 2020, equivalent to 55 billion pounds, which is approximately twice the annual production in the US and eight times the US pork exports for the same year (Speer, 2019).

ASFV is endemic in sub-Saharan Africa where it is hypothesized to have originated in warthogs but has since been established as a common virus of domestic pigs in the area. Outbreaks have occasionally occurred outside of Africa, but the virus was almost always brought under control and eradicated. One exception has been the persistence of the virus on the Italian island of Sardinia where uncontrolled movement of pigs, free-range production systems, and socioeconomic factors have complicated the situation (Gogin et al., 2013). There are more than 20 genotypes of ASFV and isolates demonstrating a full spectrum of virulence ranging from highly pathogenic strains to less virulent ones (Achenbach et al., 2017; De Villiers et al., 2010; Quembo et al., 2018). Depending on the strain, ASF can present with extremely high case-fatality, killing nearly the entire herd, or it can cause milder, nonspecific illness that can be extremely difficult to recognize. ASF only affects members of the pig family, Suidae, including domestic swine and wild boar. Wild suids, particularly warthogs, giant forest hogs, and bushpigs, serve as reservoirs and maintenance in Africa is via a complex sylvatic cycle involving transmission between the wild suids, argasid ticks, and domestic pigs (Costard et al., 2009; Dixon et al., 2019).

Even though ASFV is tick-borne, it can be effectively transmitted with or without a tick intermediate. Domestic pigs can shed virus in all secretions and excretions and, complicating recognition and control, a significant amount of shedding can occur prior to onset of clinical signs. As a high titer viremia results, massive environmental contamination is probable during necropsies, fighting between pigs, or if a pig develops melena (bloody stool). The ASF virus is resistant to many environmental conditions and persists in contaminated soil for long periods (Dixon et al., 2019). Outside of areas with the reservoir species, ASFV is most often transmitted directly between pigs via oral and nasal routes, but the pathogen can also be transmitted by iatrogenic means (Colgrove et al., 1969; Dixon et al., 2019; McVicar, 1984). Although all *Ornithodoros* species tested to date show susceptibility to infection with the virus, there appears to be varying degrees of vector competence. As a TAD, ticks play a limited role in ASF transmission when compared to the role played by movement of infected animals and animal products (EFSA, 2010).

Non-tickborne infection occurs after virus penetrates the animal's mucous membrane through direct contact with infected pigs or the environment, but most infections are thought to occur after either inhalation or ingestion. As ASFV can persist in tissues after death, significant spread can occur by feeding uncooked or undercooked, contaminated pig swill. Virus can also spread on fomites, including vehicles, feed, and equipment and the virus has been reported to persist for days to weeks in feces or urine under virus-favorable conditions (Costard et al., 2009; Dixon et al., 2019).

African swine fever causes a hemorrhagic disease in domestic pigs and can occur in varying forms from highly lethal, with case-fatality rates reaching 100%, to subclinical disease (Rock, 2017; Villeda et al., 1995; Villeda et al., 1993). It can present as peracute, acute, subacute, or chronic illness and the progression is generally correlated with the virus isolate. Sudden death without overt signs of disease (peracute presentation) can be the first sign of infection in a herd. Acute cases typically present with high fever, anorexia, lethargy, and recumbency with a reluctance to move. Pigs may have erythema and/or cyanotic blotching of the skin predominately affecting the ears, tail, and lower limbs and pigs can also present with diarrhea, constipation, vomiting, signs of abdominal pain, or hemorrhagic signs (including epistaxis and petechial/ecchymotic skin lesions). Subacute disease has similar signs with decreased severity, while chronic disease can present as an intermittent low-grade illness accompanied by depression and decreased appetite. Host and viral factors responsible for the variation in disease remain poorly understood. From a disease control perspective, it must be noted that ASFV can establish long-term persistent infections in warthogs and domestic pigs that survive acute infection (De Kock et al., 1940; DeTray, 1957; Mebus and Dardiri, 1980; Rock, 2017; Thomas et al., 2016; Wilkinson et al., 1983).

Currently, there are no effective vaccine candidates available to combat ASF. The history and progress towards the development of effective biologic countermeasures will be discussed in the next section of this chapter. Without an effective vaccine, ASF control is based on early detection, isolation, and culling of infected herds, and strict biosecurity measures to prevent introduction of the virus onto new farms. Depending on the situation, these measures can have varying success rates.

The "ideal" livestock TAD vaccine

The basis of vaccination is to simulate the development of naturally acquired immunity by inoculation of immunogenic components of a target pathogen, or similar surrogate, while not causing disease (Meeusen et al., 2007). Both veterinary and human vaccinology has made immense progress since its inception when Edward Jenner's inoculation with cowpox, a zoonotic disease, as a surrogate for prevention of smallpox disease in humans (Adams et al., 2011). This humble beginning demonstrates the close relationship between both human and animal infectious disease research, but even with the shared beginnings and close relationship, the criteria for what is considered successful vaccination and the "ideal" vaccine can be drastically different for veterinary and human biologics. Within veterinary vaccinology, there can be diverse benchmarks between target animal groups being considered. For instance, human vaccines share similar goals as companion animal ("pet") vaccination where the health and well-being of the individual animal are of chief concern. While for the livestock industry vaccination aims to safeguard and expand the economic benefit to the farmer with the main objective being to improve production of the animal and protect the herd and, therefore, maintain food security. In the case of zoonotic diseases affecting livestock, vaccination can also be used to reduce or eliminate the risk of transmission to humans involved in production and the end consumer, while also improving animal productivity. Vaccination of livestock also contributes to the welfare of animals and to the reduced need for antibiotics to treat infection (Adams et al., 2011; Meeusen et al., 2007; Roth, 2011).

The process of developing and bringing veterinary vaccines to market can be time consuming and expensive. Since most private companies are profit driven, research and development investment decisions in the veterinary industry are based on the potential of products to generate sales and recuperate costs in a given target market, this can be difficult to accomplish with many emerging infectious diseases and TADs

(Hardham and Lamichhane, 2013; Heldens et al., 2008). The potential financial returns to the manufacturer on animal biologics are typically much less than their human counterparts due to lower prices and smaller market sizes, yet their complexity and range of potential hosts and pathogens are greater. This ultimately results in lower overall investment in research and development. However, compared to human vaccine development, veterinary vaccines generally have less stringent regulatory requirements and shorter market to launch times with quicker return on investments. Animal vaccine development has the advantage of the ability to do challenge studies in the intended species. In addition, legal liabilities for manufacturers are less for animal vaccines compared to human vaccines. Aside from the safety and efficacy profiles, the predominate factor that determines the success of a vaccine is an effective commercialization plan resulting in its use in the field (Heldens et al., 2008; Meeusen et al., 2007).

The following characteristics of an "ideal" TAD vaccine are desirable, but not all are necessarily essential or currently even considered scientifically feasible:

- Shown to be pure, safe, potent, and efficacious in the target species
- Provide rapid onset with long duration of immunity with a single dose
- Are effective in the presence of maternal antibody
- Show lack of interference with other vaccines used in livestock
- Safe for manufacturing under BSL2 conditions when produced in a disease-free country
- The organism used to produce the vaccine is not considered a select agent in the U.S.
- Can be produced with cost-effective manufacturing – the cost of the product must be beneficial to the manufacturer as well as the consumer
- Are safe for use in food production animals with no, or reasonably short, withdrawal time for animal products for human consumption
- Are considered non-hazardous for humans accidentally exposed to the vaccine
- Safety and efficacy demonstrated in multiple species (where applicable)

An important consideration for TAD vaccines is the availability of a companion diagnostic assay that allows detecting infection in vaccinated animals (DIVA). This is typically accomplished by discriminating the antibody responses that are vaccine-induced from those induced during infection with the wild-type virus. For example, FMDV vaccination induces the development of antibodies to virus capsid structural proteins, whereas infection induces antibodies to both structural and non-structural proteins. The non-structural proteins are required for FMDV replication and are conserved across all serotypes and strains. To be able to utilize a DIVA diagnostic for post-vaccination surveillance, the vaccines utilized must be purified to the point that they are free of the non-structural proteins (Barnett et al., 2015b; Clavijo et al., 2004). For several TADs of livestock, effective conventional vaccines are available but their use is restricted due to their interference with disease surveillance that is dependent on determining animal infection status based on serological assays. DIVA companion diagnostics may not be required in situations when vaccinated animals are intended to go directly to slaughter as these animals are removed from the national herd. Use of non-DIVA vaccines in non-endemic countries in animals not intended for slaughter has the potential to result in loss of the country's disease-free status from the World Organization for Animal Health (OIE). FMD in cattle is a classic example of this. Even though efficacious inactivated FMD vaccines have been available for decades, they are not used in non-endemic countries as this would forfeit the country's OIE disease free without vaccination status and, therefore, compromise continued international trade (Barnett et al., 2015a, b). The following subsections will discuss more of the specifics concerning available FMD vaccines and the importance of DIVA capability in regards to eradication after introduction to a non-endemic area as well as the historical attempts at producing an efficacious vaccine for ASF and why an effective vaccine has remained elusive.

Foot-and-Mouth Disease Vaccines

Vaccination for FMDV is an essential component of planning for emergency response control measures and is also routinely used to control the disease in endemic countries. Historically, FMD vaccination has been used to successfully eliminate the disease throughout most of the western hemisphere. The presence of numerous serotypes compounds the complicated nature of vaccination as vaccination of animals with one serotype does not necessarily confer protection against other serotypes (Grubman and Baxt, 2004). In actuality, the vaccine used can even fail to provide adequate protection against other subtypes within the same serotype. Traditional FMDV vaccines are produced from live, infectious virus grown in mass quantities in cell culture that is then inactivated, purified, and adjuvanted

(Robinson et al., 2016; Rodriguez and Grubman, 2009). These traditional inactivated vaccines have fostered the successful elimination of this disease from many areas of the world and have allowed for the control and re-elimination after incursion into non-endemic areas. This formulation is the standard vaccine type for stockpiling in all of the FMD Vaccine Banks as vaccine antigen concentrate (inactivated, non-adjuvanted antigen concentrate) which can be stored over liquid nitrogen for extended periods of time and allows for rapid emergency formulation into finished vaccine (Lombard and Fussel, 2007; Robinson et al., 2016). Though these traditional vaccines have been used successfully for eradication, they have their disadvantages, including the requirement of cold chains when deploying the vaccine, the need for high-containment vaccine production facilities capable of working with infectious virus, time required to finish the concentrate into final vaccine, and the possibility of release of live or inadequately inactivated virus from manufacturing sites (Grubman and Baxt, 2004; OIE, 2013).

The concerns posed by traditional, inactivated vaccines have given rise to research into numerous novel FMDV vaccine platforms. Historically, there have been problems associated with the development of attenuated vaccines including unstable phenotypes, differences in pathogenesis between target species, and development of attenuation to the point that the virus does not consistently produce a protective immune response (Rodriguez and Grubman, 2009). Advances in cDNA technology have allowed some progress towards a phenotype stable attenuated virus, including the engineering of a serotype A12 strain lacking the leader protein coding region that demonstrated attenuation in cattle but only partially protected the animals from subsequent wild-type FMDV challenge (Brown et al., 1996; Mason et al., 1997). Inoculation of swine with the same candidate virus prevented clinically apparent disease, lack of transmission to naïve swine, and provided partial protection from wild-type challenge. When produced as a chemically inactivated and adjuvanted finished vaccine, the resulting biologic induced protection similar to that produced by finished vaccine utilizing the wild-type virus (Chinsangaram et al., 1998). Vaccines produced from viruses engineered in this fashion may provide a means of overcoming the risk of production using their highly virulent wild-type parent strains (Cao et al., 2016).

Without effective vaccination, traditional tools to control FMD in a previously FMD-free country are limited to preventing the movement of animals, heightened on-farm biosecurity, stamping out of affected populations (depopulation of all infected herds and in-contact herds as soon as possible), trace back and trace forward of cases 28 days prior to the start of the outbreak, and the rapid deployment of effective diagnostics. Other components of effective control programs are discussed later in this chapter (Barnett et al., 2015a; Grubman and Baxt, 2004).

African Swine Fever Vaccines: hurdles and progress

Despite being first described in 1921, effective vaccines protecting against ASF still remain elusive, leaving mass culling the only effective disease control option (Bellini et al., 2016; Montgomery, 1921; Rock, 2017). As ASF is one of the most significant emerging disease threats for the swine industry worldwide, development of vaccination strategies is paramount to limiting the potential devastation this disease may cause. It is thought that vaccination against this virus will be possible in the future based on knowledge that pigs that do recover from infection develop protection against reinfection with a homologous strain of the virus (King et al., 2011; Lacasta et al., 2015; Leitão et al., 2001; Mebus and Dardiri, 1980).

Typical of viral infections, both humoral and cellular immunity have been shown to be important for protection in ASFV infected pigs, but there is a lack of understanding of the specific mechanisms mediating anti-ASFV immunity and there have been no definitive correlates of protection established (Argilaguet et al., 2012; King et al., 2011; Lacasta et al., 2015; Oura et al., 2005). Viral proteins that function as protective antigens need to be identified, along with an understanding of the circulating strain variations, further understanding of viral pathogenesis, and to be able to produce broadly-effective vaccines offering cross-protection against relevant ASFV strains (Rock, 2017). As this is a large, complicated DNA virus, there are many proteins to evaluate. To be suitable for use under emergency conditions in non-endemic countries, the eventual vaccines must be DIVA compatible.

In pigs that have been naturally infected with low virulent strains of ASFV and for pigs inoculated with viruses attenuated by traditional methods, protective immunity is induced against homologous (rarely heterologous) virus challenge, but the conditions of cross protection are not well understood. Some strains produce levels of cross protection while those that appear to be closely related do not (King et al., 2011; Lacasta et al., 2015; Leitão et al., 2001; Mebus and Dardiri, 1980; Mulumba-Mfumu et al., 2016). Because of this, traditional live-attenuated vaccines are characterized by absent clinical signs and reduction in viremia and are thought to be protective due to early control of virus replication (Rock, 2017). Production processes involving live, infectious virus leads to similar hurdles as those described for FMDV, particularly regarding concerns of inadvertent release from a manufacturing facility. Attempts at engineering attenuated viruses for the purpose of vaccination has provided some progress towards a viable vaccine, but the effects of gene deletion on attenuation and immunogenicity appear to be strain dependent and inconsistent (Abrams et al., 2013; Afonso et al., 1998; Neilan et al., 2002; O'Donnell et al., 2015a; O'Donnell et al., 2015b; O'Donnell et al., 2016). Significant safety concerns have also been a major issue for live-attenuated ASFV vaccines, as shown by the first ASF vaccine study that lead to the development of chronic ASF after the originating virus had been attenuated by serial passage in bone marrow cell culture (Manso-Ribeiro et al., 1963). In general, the suitability of a live-attenuated approach to vaccination against ASFV is hindered by issues concerning efficacy, residual pathogenicity and reversion to virulence, and the development of long-term viral persistence. So far, attempts at conveying protection using a variety of inactivated vaccines have failed after numerous attempts. Future development of a vaccine using this approach appears unlikely (Abrams et al., 2013; Blome et al., 2014; Forman et al., 1982; Kihm et al., 1987; Mebus, 1988; Reis et al., 2016; Rock, 2017; Stone and Hess, 1967).

The potential use of subunit vaccines utilizing only specific protective viral antigens against ASFV within an optimized delivery or vector system may show the most promise towards a suitable vaccination approach. For these vaccines to be designed, the relevant protective antigens against ASFV and their natural antigenic diversity must be established. Results of attempts at producing an effective subunit vaccine so far indicate that additional unknown ASFV proteins are required for inducing full protective immunity (Argilaguet et al., 2013; Borca et al., 1998; Borca et al., 1994a; Borca et al., 1994b; Burmakina et al., 2016; Gomez-Puertas et al., 1996; Neilan et al., 2004; Rodríguez et al., 1993). Even though there are several viral proteins known to be correlated with protection, there is no single protein known to be sufficient for induction of sufficient protection in pigs to be utilized in an effective vaccine. This supports the conclusion that multiple viral antigens will be required for protection.

Recently, progress has been made towards the antigen selection critical for development of a subunit ASFV vaccine. Previously, using viral-vectored antigen pools, researchers tested the immunogenicity of twenty-eight different genes and demonstrated induction of antigen-specific cellular responses (Netherton et al., 2019). Goatley et al (2020) have determined that different combinations of antigens and doses could successfully induce cellular and humoral immune responses capable of whole virus recognition, yet there were no clear correlates of protection that could be used to predict outcomes after challenge with virulent ASFV. Recent work describes the induction of both antigen and ASFV-specific antibody and cellular responses with different viral-vectored pools of antigen that were selected based on their immunogenicity in pigs. One of the tested pools that consisted of eight viral-vectored genes provided 100% protection in pigs after an otherwise fatal challenge with ASFV. The surviving animals did develop a viremia as well as clinical signs post-challenge. Further studies were recommended to determine the ability of the animals to clear virus and, importantly, if the vaccinates are able to shed infectious virus capable of transmission to naïve animals (Goatley et al., 2020). Even though this study utilized genotype I antigens and challenge, other studies have suggested that there may exist some level of cross-protection between the two genotypes with possible differences in T-cell or antibody epitopes (Gallardo et al., 2018; Goatley et al., 2020; Mur et al., 2013). There is still the need to define correlates of protection, the requirement for reimmunization with booster vaccination, and if other antigens may be utilized to improve the efficacy of the vaccine (Goatley et al., 2020). This work further validates the attractiveness of viral vector vaccine platforms as a potential mechanism for ASF control and continues to provide encouragement that there will one day be successful countermeasures to combat this terrible disease. Even with the significant amount of progress and expansion of knowledge concerning ASFV and the disease that it causes, there still remain significant challenges before an effective vaccine becomes a reality and can be incorporated into disease control measures.

Nipah virus

Nipah virus (NiV) is a member of the family *Paramyxoviridae*, genus *Henipavirus* that emerged in dramatic fashion in Malaysia and Singapore in 1998, initially being mistaken for an atypical presentation of another viral disease, Japanese encephalitis. Nipah virus-induced disease in humans was characterized by severe febrile encephalitis that led to a case fatality rate of approximately 40% during the first outbreak. Although *Pteropus* bats are considered to be the predominant natural reservoir of the virus, human infections in Malaysia were found to be from zoonotic transmission from domestic pigs (AbuBakar et al., 2004; Chua et al., 2002; Clayton et al., 2013). Even with a quick national response and international assistance, the first outbreak ultimately led to the death of 105 people and the mass culling of almost half of the country's national pig herd, approximately 1.1 million head were lost. NiV is also suspected to be the causative agent of an outbreak in horses in the Philippines with subsequent transmission to humans (Ching et al., 2015). Since 2001, a different genotype of NiV has emerged with near annual outbreaks occurring in Bangladesh where sustained human to human transmission is responsible for continued spread of the disease with a case fatality rate reaching above 75%. In the case of this virus, NiV-Bangladesh, transmission is associated with bat urine-contaminated date palm sap that is consumed by people (Broder et al., 2016; Homaira et al., 2010a; Homaira et al., 2010b; Luby et al., 2009; Rahman et al., 2012).

Natural infection of pigs can present with a spectrum of clinical signs, ranging from asymptomatic to mild-moderate respiratory diseases and, occasionally, encephalitis as part of a "porcine respiratory and encephalitis syndrome" (Nor et al., 2000). These mild, non-specific clinical presentations in pigs are problematic as they can easily go unnoticed until consequent human infections begin to surface.

Nipah virus, as recognized by the Coalition of Epidemic Preparedness Innovations (CEPI), is one of the viral threats of greatest concern for a human pandemic. It has also been proposed as one of the priority diseases of interest to the United States Department of Agriculture and the Department of Homeland Security, as well as being listed as a US Public Health Service Category A agent, due to its potential for use for intentional release as a component of bioterrorism. Due to the susceptibility of humans, high case-fatality rates for NiV disease, absence of human vaccines or therapeutics, this virus is classified as a Biosafety Level 4 (BSL4) pathogen (Chosewood and Wilson, 2009). Vaccine development, production, and efficacy testing are made even more complex due to the heightened regulation and strict constraints for biosafety and containment when working with this classification of pathogens, ultimately leading to a reluctance for the biologics industry to invest in products without government support, culminating in delays in research and in bringing an effective vaccine to market (Hardham and Lamichhane, 2013)

CEPI has allocated $460 million (USD) for development, licensing, and stockpiling of human vaccines for what they have determined are the three highest pandemic threat agents (Nipah virus, Lassa Fever virus, and Middle Eastern Respiratory Syndrome coronavirus), this includes no provisions for prevention or control of Nipah virus in swine (Sifferlin, 2017). Due to the nature and consequences of NiV virus outbreaks in Southeast and South Asia, as well as the potential for unintentional or intentional introduction into other areas of the world, a swine vaccine able to eliminate viral shedding is of paramount importance and should be considered as an important control measure in pigs to ultimately control disease emergence in humans. If NiV should emerge or reemerge in a swine dense region, a highly efficacious vaccine for use in domestic swine would serve as a critical protective measure for both human and animal health, as well as a means of protecting the affected nation's food supply and economy. If cost-effective and available, a vaccine could also serve to protect small-holder investments in swine and maintaining food security, which is critical in the areas of likely reemergence.

The foundation of scientific knowledge based on many years of experimental work with NiV has provided the framework that should support successful development of a domestic swine vaccine. Numerous active and passive immunization approaches have been explored in a variety of species, as reviewed by Broder et al, 2016, including live-recombinant virus platforms, protein subunit, virus-like particles, and DNA vaccines (Broder et al., 2016). Most of these attempts have been focused on the development of NiV-specific neutralizing antibodies, however, some have evaluated both the ability to illicit an immune response as well as efficacy against challenge utilizing pigs. Initially, use of a recombinant canarypox virus with NiV F and G glycoprotein genes inserts was evaluated in piglets with vaccination followed by intranasal challenge with live NiV. This approach demonstrated that prevention of

NiV-mediated illness was achievable with minimal detection of viral nucleic acid being shed and, critically, live virus was not recoverable from vaccinated animals (Broder et al., 2016; Weingartl et al., 2006). It was then demonstrated that both arms of the immune response, cell-mediated and humoral, are critical for the protection of domestic swine against NiV infection (Pickering et al., 2016). Despite the promising results with the canarypox vector platform in swine and the numerous platforms evaluated in other species, a swine vaccine for NiV is still not available even with the continued threat of new spillover into this widespread species (Broder et al., 2016).

The desirable characteristics of a NiV vaccine to aid in the control of an outbreak in a swine dense region should be considered similar to other vaccines for use in combating transboundary animal diseases. There must be rapid onset of immunity, ideally through a single dose, and the induced immunity should be protective and long lasting. The vaccine should not interfere with booster vaccination or other vaccines being used in swine for endemic diseases and it should induce effective immunity in the face of maternal antibody so that it can be utilized in young animals. It should be safe to manufacture in BSL2 conditions. There should be a companion diagnostic assay available to determine infection versus vaccination status (DIVA). The product should be safe for use in food producing animals with a reasonably short withdrawal period for products for human consumption. Importantly, the vaccine should not be hazardous for humans that may be accidently exposed. To have sufficient product available to rapidly respond to an outbreak in swine that is either too large and/or too widespread to effectively stamp out through culling, a safe and efficacious NiV vaccine must be a component of a national or international veterinary stockpile. These countermeasures could be finished vaccine that is ready to deploy, stocks of vaccine antigen concentrate which can be readily converted into finished vaccine, and there must be the capability of rapidly producing large volumes of new vaccine.

Stockpiling vaccines for emergency response

Vaccine banks are typically stock of vaccine antigen concentrate, ready to be formulated into finished vaccines or ready-to-use products able to be rapidly deployed for emergency response with the overall goal of reducing the time between detection of disease and vaccination response. Vaccine banks were historically developed by FMD-free countries to control unexpected incursion of disease after vaccination campaigns end and disease-free status is gained (Barnett et al., 2010; Lei and McKercher, 1979; Lombard and Fussel, 2007). In fact, the first mention of this approach was following a devastating FMD outbreak in Great Britain in 1967-1968 after a commission charged with examining the outbreak made the recommendation to enhance preparedness for future control. Following the commission report, the British government purchased finished vaccine annually to cover several hundred thousand head of cattle for serotypes O, A, and C (Lombard and Fussel, 2007). This established the first known strategic antigen bank. Since this was finished, fully formulated vaccine, the shelf-life of the product required frequent replacement. In the 1970s, manufacturers developed novel methods to purify, concentrate, and store inactivated FMD antigen that allowed for resistance to freezing at ultra-low temperatures (liquid nitrogen). This process ultimately led to an extension of shelf-life when product was stored as vaccine antigen concentrate (VAC) (Barnett et al., 2010; Lombard and Fussel, 2007). A major drawback for the storage of VAC is the time required to take the VAC and finish (blending, formulating, filling) into a final vaccine that was ready for field deployment. Banks were principally established to combat inadvertent incursions of disease, but, in recent years, there has been an increased recognition of the possible consequences of deliberate use in agroterrorism (Barnett et al., 2010).

Regardless of the diseases to which the countermeasures are maintained, vaccine banks face many issues including strain or serotype selection, as well as the regulation of the manufacturing, storage, formulation, and testing of stored products, security, maintenance, monitoring, and disposal of products (Barnett et al., 2010). This regulation is even more complex when banks represent multiple countries, as each countries regulation and licensing requirements typically differ. Most vaccines utilized to combat TADs are produced for endemic country use and are made to order for a specific need by their manufacturer and provided formulated and ready to use. This typically occurs without surplus product and provides little flexibility to accommodate revisions to specifications at short notice (Barnett et al., 2010). This, of course, emphasizes the need for the vaccine banks. There are shared, desirable characteristics of vaccines and VAC meant for emergency use and kept in strategic reserves that should be considered when stockpiling a vaccine:

- The vaccine should meet as many of the desirable characteristics for that agent as possible.
- There should be ongoing manufacture and sale of the products in endemic countries. This enables indefinite delivery and indefinite quantity contracts for just-in-time delivery, reducing the need for storage in traditional banks.
- The product should be based on antigen that is stable when stored as bulk VAC.
- Finished vaccine should have an established long shelf-life and stability of the finished product.
- The finished vaccine should be heat stable to reduce the need for extensive cold-chain management during deployment.
- There should be rapid scale up and manufacture processes in the regions of concern.

Vaccination as a component of an effective Control Program

This chapter, thus far, has focused on the need for efficacious vaccines for TADs, but vaccination is only one part of what is required for an effective Control Program in response to disease incursion. These programs must be designed and carried out by the competent veterinary authority for specific countries or regions, as there is no "one size fits all" response and every outbreak is different. Even with these differences, there are similar, overarching components that include the following:

- Biosecurity
- Surveillance and the ability to trace-back and trace-forward from known cases, to include:
  - Animal identification
  - Premises identification
  - Movement records
- Vaccination protocols (species, age, frequency, monitoring of the development of immunity, matching of vaccine to the outbreak strain; for example, with FMDV)
- Vaccine quality control from production through administration (including cold chain maintenance, proper administration, records)
- The need to understand the characteristics of the vaccine:
  - Expected efficacy in a naïve population
  - Efficacy in the presence of maternal antibody

  - Duration of immunity
- Diagnostic assays capable of use for post-vaccination surveillance (DIVA)
- Ultimately, the ability to suppress transmission (the “reproduction ratio”) in a population.

Eurasian African Swine Fever Outbreak: uncontrolled spread of a TAD due to lack of a vaccine

ASF emerged in the country of Georgia in 2007 where it was hypothesized that the virus
was introduced from contaminated pork products originating from a ship being fed to pigs in the region of the Port of Poti (Rowlands et al., 2008). After rapid spread involving both domestic pigs and wild boar into Armenia, Azerbaijan, and Iran, the outbreak reached the Russian Federation in early 2008. Since 2014, the virus has also been circulating in eastern Europe, including Estonia, Latvia, Lithuania, and Poland. The incursion in the Russian Federation brought the virus into a bordering country, putting the outbreak in close proximity to the People's Republic of China (Gogin et al., 2013; Sánchez-Vizcaíno et al., 2013).

Pig production and pork consumption are important to China for cultural and economic reasons as over 500 million pigs are produced within its borders annually, representing over half of the global pig population (Vergne et al., 2017). The proportion of pigs being produced in small, backyard farms raising less than 30 sows is estimated to be around 40% of the country's total production (Gilbert et al., 2015; McOrist et al., 2011). The factors that were proposed to put China at greatest risk for ASF introduction were discussed by Vergne et al, 2017. The risks specific to China included (1) the international movements of people in and out of the country and the potential for carrying contaminated or infected goods, (2) the significant increase in the demand for pork leading to large volumes of pork and pork products being imported into the country – both legally and illegally, (3) the widely distributed practice of raw swill feeding to subsidize and support pig production, and (4) the widespread population of wild boar throughout most of the country (Vergne et al., 2017).

In August, 2018, an incursion of ASF into Shenyang in northeastern China quickly spiraled into an unprecedented disaster and challenge to the Chinese swine industry

(Wang et al., 2018; Zhou et al., 2018). The sequence of the circulating outbreak strain shares high homology with the Georgia 2007/1 strain and it is suggested that introduction occurred via smuggled pork originating from an Eastern European country (Ge et al., 2018; Wang et al., 2018). Since its introduction, ASF has been confirmed in numerous provinces across the country (Figure 3). The complexity of pork production and the fact that a major proportion of China's pigs are raised in smallholder farmyards with minimal biosecurity makes epidemiologic tracing a monumental task, but investigators have proposed that the virus has been circulating in country since at least March, 2018 (FAO, 2020; Normile, 2018).

Without a vaccine, control of ASF relies on early detection and rapid, local eradication efforts (Wang et al., 2018). The governmental response has been strong, involving the culling of sick and exposed animals; isolating infected farms; controlling the movement of live pigs and pig products; conducting extensive epidemiologic tracing; and disinfecting farms, processing facilities, and markets (Normile, 2018). Importantly, this response has also included trying to earn the cooperation of producers through immediate compensation for culled animals with the ultimate goal of preventing the slaughter of sick pigs and the sale and transport of contaminated meat that could perpetuate the continued spread of the virus (Normile, 2018; Wang et al., 2018). Even with this tremendous effort, ASF has spread rapidly throughout China and into neighboring countries (FAO, 2020; Zhao et al., 2019).

Researchers found that the particular ASF virus causing the outbreak in China is highly virulent, causing acute disease in 3-5 days in experimentally inoculated pigs with death occurring within 9 days and acute disease in contact animals in 9 days with death occurring within two weeks (Zhao et al., 2019). Further complicating the situation, a novel variant of the virus has begun spreading in the wild boar population representing the possibility of the introduction of a new strain into China or the evolution of a pre-existing strain (Li et al., 2019). This will only serve to further complicate the situation, as well as any potential actions taken towards eradication.

Given the massive scale of the Chinese pork sector, the impact of ASF has the potential to be devastating. Introduction of this TAD has put a crucial protein source at risk and caused income insecurity for many families that are dependent on their household farms. There are many less obvious, indirect consequences of the

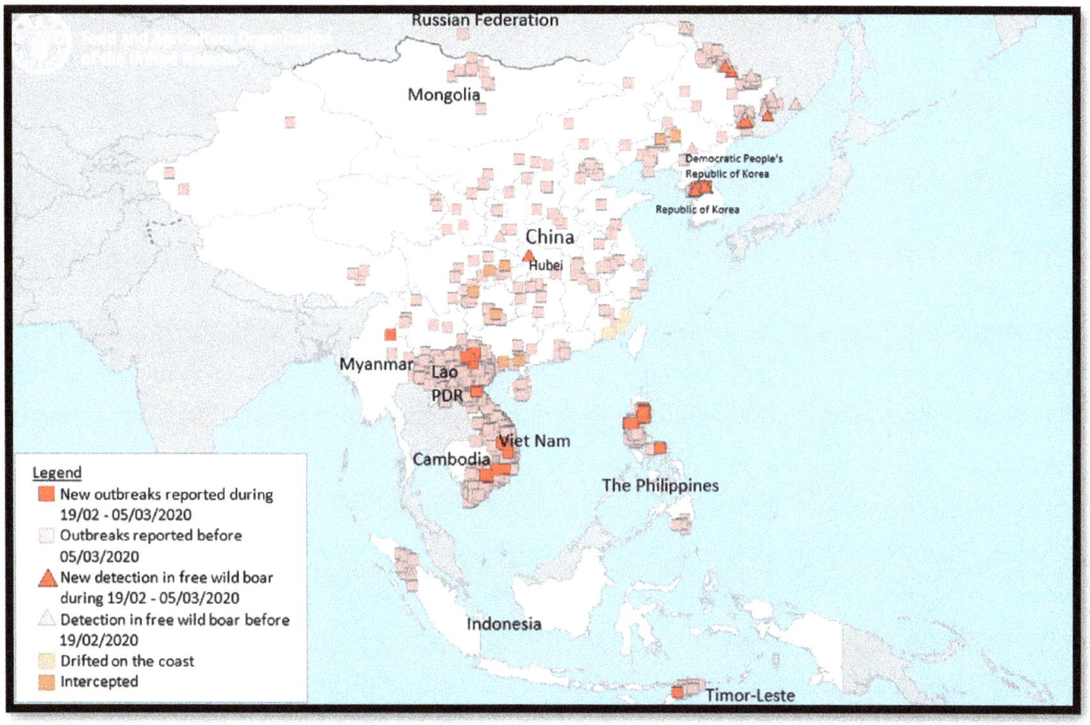

Figure 3. A map of the People's Republic of China showing the widespread distribution of African swine fever virus across numerous Chinese provinces and the spread to neighboring countries. From FAO, 2020.

outbreak that are beginning to be noticed. For instance, most of the unfractionated and low-molecular weight heparins used worldwide are produced in China from porcine mucosa. This outbreak and the consequent losses to the Chinese pig herd are jeopardizing the global availability of this critically important pharmaceutical (Vilanova et al., 2019).

Though various vaccines against ASF are being developed, there are no commercial products currently available for use (Wang et al., 2018). A safe and effective product must be considered a priority for cost-effective control measures to be a possibility in areas with similarly extensive swine populations. Emergence of ASF in Chinese pigs is an unprecedented disaster for China, as well as the rest of the world pork producers. If this virus remains in China, it will continue to pose a major threat to the rest of the world. ASF is a formidable TAD and its threat must not be taken lightly. All of these factors have set the stage for a costly, long-lasting, and challenging outbreak that will involve significant socioeconomic impacts and investment by the associated governments involved and the swine industry as a whole.

Rinderpest: a complex interplay between animal and human health and a vaccination success story

Rinderpest is an example of how a TAD that only affects animals can have a devastating impact on human health, but can ultimately be brought under control through the successful use of efficacious vaccines. Rinderpest virus is a member of the genus *Morbillivirus* with only one serotype, but three genetically distinct lineages that were found in different geographic areas. This virus causes acute, highly contagious disease of cattle, domesticated buffalo, and a variety of species of wildlife. The impact of Rinderpest led to the establishment of the first veterinary school in France in 1762 with the direct purpose of training specialists in the disease (Gilbert, 2009). In the 1880s, infected cattle originating in India introduced the virus to the African continent via the port of Massawa, Eritrea. This led to the death of 90% of the cattle in sub-Saharan Africa with widespread death in sheep and goats, as well as decimating many wildlife species. Mass human starvation ensued after the death of domestic livestock and draft animals, killing a third of the human population in Ethiopia and two-thirds of the Maasai people of Tanzania. With the drastic reduction in numbers of grazing animals, thickets formed in the savannah that served as breeding grounds for tsetse flies, the predominant vector for human African sleeping sickness (Moutou, 2014; Normile, 2008; Roth, 2011). The Great Ethiopian famine was not caused by a shortage of meat, it was caused by a shortage of grain secondary to the reduced draught power and reduced availability of manure caused by rinderpest (Pankhurst, 1966).

The first tissue culture produced vaccine against rinderpest was developed in 1960 (Plowright and Ferris, 1962). This vaccine was a success in that it conferred lifelong protective immunity to all lineages of the virus and had no reported adverse reactions, as well as being relatively inexpensive and easy to mass produce. Unfortunately, this vaccine required strict maintenance of cold chain, requiring logistically difficult refrigeration that is especially limiting in remote areas of Africa. This obstacle was overcome in 1981 by the modification of the lyophilization process of the same vaccine that resulted in a thermostable product with a shelf-life of 8 months at 37°C, which was sufficient to recommend its use in the field for up to 30 days without refrigeration. This allowed the vaccine to be delivered to remote areas and marginalized communities (Mariner et al., 2012). Although this improved vaccine continued to confer life-long immunity, it posed a challenge as it is impossible to differentiate between infected and vaccinated animals, complicating the overall goal of eradicating the disease. Due to this, the only option for monitoring the success of the local vaccination campaign was to halt the use of vaccine locally and monitor the area for resurgence of the virus. If the disease resurfaced, the campaign had to be reinstated (Gilbert, 2009).

The Global Rinderpest Eradication Program was established 1994 by the Food and Agriculture Organization of the United Nations (FAO) and the World Organization for Animal Health (OIE) as a large-scale international collaboration charged with coordination of vaccination campaigns, local and international trade restrictions, and disease surveillance. This new charge incorporated earlier, regional efforts with a focus on widespread vaccination and long-term, epidemiologic monitoring (Gilbert, 2009). After seventeen years, rinderpest was declared eradicated by both the FAO and the OIE in May, 2011 (Mariner et al., 2012; Njeumi et al., 2012). Rinderpest is only the second disease, following smallpox in humans, to be successfully eradicated through a coordinated, international campaign. This effort is one of veterinary medicine's greatest achievements and would not have been possible without the availability of an efficacious vaccine.

Conclusion

The global human population is growing at an exponential rate leading to the need for continued expansion of intensive food animal production to meet the world's increasing nutritional requirements. As a consequence of this increased production demand, the use of high volume, animal dense systems have expanded providing high quality protein at reduced costs. However, it also provides the opportunity for the rapid spread of disease, increased fitness pressure leading to mutation and adaptation of pathogens to new host species or increased disease severity in historically susceptible populations. This continues to drive the emergence and reemergence of diseases. The increase in "backyard" animal production in both rural and urban areas provides a critical source of high-quality protein and income to many, while also providing further opportunity for uncontrolled spread of disease and increasing the difficulty in pathogen control measures. The increase in all types of animal production to feed the increasing human population increases the urgency for the development and availability of efficacious vaccines for transboundary animal diseases

References

Abrams, C.C., Goatley, L., Fishbourne, E., Chapman, D., Cooke, L., Oura, C.A., Netherton,C.L., Takamatsu, H.-H., Dixon, L.K. (2013). Deletion of virulence associated genes from attenuated African swine fever virus isolate OUR T88/3 decreases its ability to protect against challenge with virulent virus. Virology 443, 99-105.

AbuBakar, S., Chang, L.-Y., Ali, A.M., Sharifah, S., Yusoff, K., Zamrod, Z. (2004). Isolation and molecular identification of Nipah virus from pigs. Emerging infectious diseases 10, 2228.

Achenbach, J., Gallardo, C., Nieto-Pelegrín, E., Rivera-Arroyo, B., Degefa-Negi, T., Arias, M., Jenberie, S., Mulisa, D., Gizaw, D., Gelaye, E. (2017). Identification of a new genotype of African swine fever virus in domestic pigs from Ethiopia. Transboundary emerging diseases 64, 1393-1404.

Adams, L.G., Khare, S., Lawhon, S.D., Rossetti, C.A., Lewin, H.A., Lipton, M.S., Turse, J.E., Wylie, D.C., Bai, Y., Drake, K.L. (2011). Enhancing the role of veterinary vaccines reducing zoonotic diseases of humans: linking systems biology with vaccine development. Vaccine 29, 7197-7206.

Afonso, C., Zsak, L., Carrillo, C., Borca, M., Rock, D. (1998). African swine fever virus NL gene is not required for virus virulence. General Virology 79, 2543-2547.

Alexandrov, T., Stefanov, D., Kamenov, P., Miteva, A., Khomenko, S., Sumption, K., Meyer-Gerbaulet, H., Depner, K. (2013). Surveillance of foot-and-mouth disease (FMD) in susceptible wildlife and domestic ungulates in Southeast of Bulgaria following a FMD case in wild boar. Veterinary microbiology 166, 84-90.

Amass, S.F., Mason, P.W., Pacheco, J.M., Miller, C.A., Ramirez, A., Clark, L.K., Ragland, D., Schneider, J.L., Kenyon, S.J. (2004). Procedures for preventing transmission of foot-and-mouth disease virus (O/TAW/97) by people. Veterinary microbiology 103, 143-149.

Argilaguet, J.M., Pérez-Martín, E., López, S., Goethe, M., Escribano, J., Giesow, K., Keil, G.M., Rodríguez, F. (2013). BacMam immunization partially protects pigs against sublethal challenge with African swine fever virus. Antiviral research 98, 61-65.

Argilaguet, J.M., Pérez-Martín, E., Nofrarías, M., Gallardo, C., Accensi, F., Lacasta, A., Mora, M., Ballester, M., Galindo-Cardiel, I., López-Soria, S. (2012). DNA vaccination partially protects against African swine fever virus lethal challenge in the absence of antibodies. PloS one 7.

Arzt, J., Baxt, B., Grubman, M., Jackson, T., Juleff, N., Rhyan, J., Rieder, E., Waters, R., Rodriguez, L. (2011a). The Pathogenesis of Foot-and-Mouth Disease II: Viral Pathways in Swine, Small Ruminants, and Wildlife; Myotropism, Chronic Syndromes, and Molecular Virus–Host Interactions. Transboundary and emerging diseases 58, 305-326.

Arzt, J., Juleff, N., Zhang, Z., Rodriguez, L. (2011b). The pathogenesis of foot-and-mouth disease I: viral pathways in cattle. Transboundary and emerging diseases 58, 291-304.

Bachrach, H.L. (1968). Foot-and-mouth disease. Annual Reviews in Microbiology 22, 201-244.

Barnett, P., Bashiruddin, J., Hammond, J., Geale, D., Paton, D., 2010. Toward a global foot and mouth disease vaccine bank network. Revue scientifique et technique 29, 593.

Barnett, P., Geale, D., Clarke, G., Davis, J., Kasari, T. (2015a). A review of OIE country status recovery using vaccinate-to-live versus vaccinate-to-die foot-and-mouth disease response policies I: Benefits of higher potency vaccines and associated NSP DIVA test systems in post-outbreak surveillance. Transboundary and emerging diseases 62, 367-387.

Barnett, P., Geale, D., Clarke, G., Davis, J., Kasari, T. (2015b). A review of OIE country status recovery using vaccinate-to-live versus vaccinate-to-die foot-and-mouth disease response policies I: Benefits of higher potency vaccines and associated NSP DIVA test systems in post-outbreak surveillance. Transboundary emerging diseases 62, 367-387.

Bartley, L., Donnelly, C., Anderson, R. (2002). Review of foot-and mouth disease virus survival in animal excretions and on fomites (British Medical Journal Publishing Group).

Bellini, S., Rutili, D., Guberti, V. (2016). Preventive measures aimed at minimizing the risk of African swine fever virus spread in pig farming systems. Acta Veterinaria Scandinavica 58, 82.

Blome, S., Gabriel, C., Beer, M. (2014). Modern adjuvants do not enhance the efficacy of an inactivated African swine fever virus vaccine preparation. Vaccine 32, 3879-3882.

Borca, M., Carrillo, C., Zsak, L., Laegreid, W., Kutish, G., Neilan, J., Burrage, T., Rock, D. (1998). Deletion of a CD2-like gene, 8-DR, from African swine fever virus affects viral infection in domestic swine. Virology 72, 2881-2889.

Borca, M., Irusta, P., Carrillo, C., Afonso, C., Burrage, T., Rock, D.L. (1994a). African swine fever virus structural protein p72 contains a conformational neutralizing epitope. Virology 201, 413-418.

Borca, M., Kutish, G., Afonso, C., Irusta, P., Carrillo, C., Brun, A., Sussman, M., Rock, D.L. (1994b). An African swine fever virus gene with similarity to the T-lymphocyte surface antigen CD2 mediates hemadsorption. Virology 199, 463-468.

Brito, B., Rodriguez, L., Hammond, J., Pinto, J., Perez, A. (2017). Review of the global distribution of foot-and-mouth disease virus from 2007 to 2014. Transboundary and emerging diseases 64, 316-332.

Broder, C.C., Weir, D.L., Reid, P.A., 2016. Hendra virus and Nipah virus animal vaccines. Vaccine 34, 3525-3534.

Brown, C., Piccone, M., Mason, P., McKenna, T., Grubman, M. (1996). Pathogenesis of wild-type and leaderless foot-and-mouth disease virus in cattle. Virology 70, 5638-5641.

Burmakina, G., Malogolovkin, A., Tulman, E., Zsak, L., Delhon, G., Diel, D., Shobogorov, N., Morgunov, Y.P., Morgunov, S.Y., Kutish, G. (2016). African swine fever virus serotype-specific proteins are significant protective antigens for African swine fever. General Virology 97, 1670-1675.

Cao, Y., Lu, Z., Liu, Z., 2016. Foot-and-mouth disease vaccines: progress and problems. Expert Review of Vaccines 15, 783-789.

Chapman, D.A., Tcherepanov, V., Upton, C., Dixon, L.K. (2008). Comparison of the genome sequences of non-pathogenic and pathogenic African swine fever virus isolates. General Virology 89, 397-408.

Ching, P.K.G., de Los Reyes, V.C., Sucaldito, M.N., Tayag, E., Columna-Vingno, A.B., Malbas, F.F., Jr., Bolo, G.C., Jr., Sejvar, J.J., Eagles, D., Playford, G., Dueger, E., Kaku, Y., Morikawa, S., Kuroda, M., Marsh, G.A., McCullough, S., Foxwell, A.R. (2015). Outbreak of henipavirus infection, Philippines, 2014. Emerging Infectious Diseases 21, 328-331.

Chinsangaram, J., Mason, P.W., Grubman, M.J. (1998). Protection of swine by live and inactivated vaccines prepared from a leader proteinase-deficient serotype A12 foot-and-mouth disease virus. Vaccine 16, 1516-1522.

Chosewood, L.C., Wilson, D.E. (2009). Biosafety in microbiological and biomedical laboratories. US Department of Health and Human Services, Public Health Service, Centers ....

Chua, K.B., Koh, C.L., Hooi, P.S., Wee, K.F., Khong, J.H., Chua, B.H., Chan, Y.P., Lim, M.E., Lam, S.K. (2002). Isolation of Nipah virus from Malaysian Island flying-foxes. Microbes and infection 4, 145-151.

Clavijo, A., Wright, P., Kitching, P. (2004). Developments in diagnostic techniques for differentiating infection from vaccination in foot-and-mouth disease. The Veterinary Journal 167, 9-22.

Clayton, B., Wang, L., Marsh, G. (2013). Henipaviruses: an updated review focusing on the pteropid reservoir and features of transmission. Zoonoses and public health 60, 69-83.

Colgrove, G., Haelterman, E., Coggins, L. (1969). Pathogenesis of African swine fever in young pigs. American Journal of Veterinary Research, 1343-1359.

Costard, S., Wieland, B., De Glanville, W., Jori, F., Rowlands, R., Vosloo, W., Roger, F., Pfeiffer, D.U., Dixon (2009). African swine fever: how can global spread be prevented? 364, 2683-2696.

De Kock, G., Robinson, E., Keppel, J. (1940). Swine fever in South Africa. Onderstepoort Journal of Veterinary Research 14.

De Villiers, E.P., Gallardo, C., Arias, M., Da Silva, M., Upton, C., Martin, R., Bishop, R.P. (2010). Phylogenomic analysis of 11 complete African swine fever virus genome sequences. Virology 400, 128-136.

DeTray, D. (1957). Persistence of viremia and immunity in African swine fever. American journal of veterinary research 18, 811-816.

Dixon, L.K., Chapman, D.A., Netherton, C.L., Upton, C. (2013). African swine fever virus replication and genomics. Virus research 173, 3-14.

Dixon, L.K., Escribano, J., Martins, C., Rock, D.L., Salas, M., Wilkinson, P.J. (2005). Asfarviridae. virus taxonomy, 135-143.

Dixon, L.K., Sun, H., Roberts, H. (2019). African swine fever. Antiviral Research 165, 34-41.

EFSA (2010). Scientific opinion on African swine fever. European Food Safety Authority 9, 1-149.

FAO (1996). Prevention and Control of Transboundary Animal Diseases. Report of the FAO Expert Consultation on the Emergency Prevention System (EMPRES) for Transboundary Animal and Plant Pests and Diseases (Livestock Diseases Programme) including the Blueprint for Global Rinderpest Eradication, FAO Animal Production and Health Paper 133, Rome, Italy, 24–26 July.

FAO (2020). ASF Situation in Asia Update (Food and Agriclutre Organization of the United Nations).

FAO (2009). High Level Expert Forum—How to Feed the World in 2050, Economic and Social Development (Food and Agricultural Organization of the United Nations Rome).

Forman, A., Wardley, R., Wilkinson, P. (1982). The immunological response of pigs and guinea pigs to antigens of African swine fever virus. Archives of virology 74, 91-100.

Gallardo, C., Sánchez, E.G., Pérez-Núñez, D., Nogal, M., de León, P., Carrascosa, Á.L., Nieto, R., Soler, A., Arias, M.L., Revilla, Y. (2018). African swine fever virus (ASFV) protection mediated by NH/P68 and NH/P68 recombinant live-attenuated viruses. Vaccine 36, 2694-2704.

Ge, S., Li, J., Fan, X., Liu, F., Li, L., Wang, Q., Ren, W., Bao, J., Liu, C., Wang, H. (2018). Molecular characterization of African swine fever virus, China, 2018. Emerging infectious diseases 24, 2131.

Gilbert, M., Conchedda, G., Van Boeckel, T.P., Cinardi, G., Linard, C., Nicolas, G., Thanapongtharm, W., D'Aietti, L., Wint, W., Newman, S.H. (2015). Income disparities and the global distribution of intensively farmed chicken and pigs. PLoS One 10.

Gilbert, N. (2009). Cattle disease faces total wipeout; rinderpest goes the way of smallpox. Nature 462, 709-710.

Goatley, L.C., Reis, A.L., Portugal, R., Goldswain, H., Shimmon, G.L., Hargreaves, Z., Ho, C.-S., Montoya, M., Sánchez-Cordón, P.J., Taylor, G. (2020). A Pool of Eight Virally Vectored African Swine Fever Antigens Protect Pigs Against Fatal Disease. Vaccines 8, 234.

Gogin, A., Gerasimov, V., Malogolovkin, A., Kolbasov, D. (2013). African swine fever in the North Caucasus region and the Russian Federation in years 2007–2012. Virus research 173, 198-203.

Gomez-Puertas, P., Rodriguez, F., Oviedo, J.M., Ramiro-Ibanez, F., Ruiz-Gonzalvo, F., Alonso, C., Escribano, J.M. (1996). Neutralizing antibodies to different proteins of African swine fever virus inhibit both virus attachment and internalization. virology 70, 5689-5694.

Grubman, M.J., Baxt, B. (2004). Foot-and-mouth disease. Clinical microbiology reviews 17, 465-493.

Hardham, J., Lamichhane, C. (2013). Biologics Industry Challenges for Developing Diagnostic Tests for the National Veterinary Stockpile, In: Vaccines and Diagnostics for Transboundary Animal Diseases. Karger Publishers, 39-52.

Hassell, J.M., Begon, M., Ward, M.J., Fèvre, E.M. (2017). Urbanization and disease emergence: dynamics at the wildlife–livestock–human interface. Trends in ecology evolution 32, 55-67.

Haydon, D., Samuel, A., Knowles, N. (2001). The generation and persistence of genetic variation in foot-and-mouth disease virus. Preventive Veterinary Medicine 51, 111-124.

Heldens, J., Patel, J., Chanter, N., Ten Thij, G., Gravendijck, M., Schijns, V., Langen, A., Schetters, T.P. (2008). Veterinary vaccine development from an industrial perspective. The Veterinary Journal 178, 7-20.

Homaira, N., Rahman, M., Hossain, M.J., Epstein, J.H., Sultana, R., Khan, M.S.U., Podder, G., Nahar, K., Ahmed, B., Gurley, E.S., Daszak, P., Lipkin, W.I., Rollin, P.E., Comer, J.A., Ksiazek, T.G., Luby, S.P. (2010a). Nipah virus outbreak with person-to-person transmission in a district of Bangladesh, 2007. Epidemiology and Infection 138, 1630-1636.

Homaira, N., Rahman, M., Hossain, M.J., Nahar, N., Khan, R., Rahman, M., Podder, G., Nahar, K., Khan, D., Gurley, E.S. (2010b). Cluster of Nipah virus infection, Kushtia district, Bangladesh, 2007. PLoS One 5.

Jamal, S.M., Belsham, G.J. (2013). Foot-and-mouth disease: past, present and future. Veterinary Research 44, 116.

Jones, B.A., Grace, D., Kock, R., Alonso, S., Rushton, J., Said, M.Y., McKeever, D., Mutua, F., Young, J., McDermott, J. (2013). Zoonosis emergence linked to agricultural intensification and environmental change. Proceedings of the National Academy of Sciences 110, 8399-8404.

Kihm, U., Ackermann, M., Mueller, H., Pool, R. 1987. Approaches to vaccination, In: African swine fever. Springer, 127-144.

King, D., Di Nardo, A., Henstock, M. (2018). OIE/FAO Foot-and-Mouth Disease Reference Laboratory Network: Annual Report 2018. In The Pirbright Institute: Woking, UK.

King, K., Chapman, D., Argilaguet, J.M., Fishbourne, E., Hutet, E., Cariolet, R., Hutchings, G., Oura, C.A., Netherton, C.L., Moffat, K. (2011). Protection of European domestic pigs from virulent African isolates of African swine fever virus by experimental immunisation. Vaccine 29, 4593-4600.

Knight-Jones, T., Rushton, J. (2013). The economic impacts of foot and mouth disease–What are they, how big are they and where do they occur? Preventive veterinary medicine 112, 161-173.

Knight-Jones, T.J.D., Robinson, L., Charleston, B., Rodriguez, L.L., Gay, C.G., Sumption, K.J., Vosloo, W. (2016). Global Foot-and-Mouth Disease Research Update and Gap Analysis: 2 – Epidemiology, Wildlife and Economics. Transboundary and Emerging Diseases 63, 14-29.

Lacasta, A., Monteagudo, P.L., Jiménez-Marín, Á., Accensi, F., Ballester, M., Argilaguet, J., Galindo-Cardiel, I., Segalés, J., Salas, M.L., Domínguez, J. (2015). Live attenuated African swine fever viruses as ideal tools to dissect the mechanisms involved in viral pathogenesis and immune protection. Veterinary research 46, 135.

Lee, S.-Y., Lee, Y.-J., Kim, R.-H., Park, J.-N., Park, M.-E., Ko, M.-K., Choi, J.-H., Chu, J.-Q., Lee, K.-N., Kim, S.-M. (2017). Rapid engineering of foot-and-mouth disease vaccine and challenge viruses. Virology 91, e00155-00117.

Lei, J., McKercher, P. (1979). Report of the 1979 Session of the Research Group of the Standing Technical Committee of EUFMD, 12-14 June, Lindholm, Denmark. Appendix B1. Food and Agriculture Organization of the United Nations, Rome 24.

Leitão, A., Cartaxeiro, C., Coelho, R., Cruz, B., Parkhouse, R., Portugal, F.C., Vigário, J.D., Martins, C.L. (2001). The non-haemadsorbing African swine fever virus isolate ASFV/NH/P68 provides a model for defining the protective anti-virus immune response. General Virology 82, 513-523.

Li, L., Ren, Z., Wang, Q., Ge, S., Liu, Y., Liu, C., Liu, F., Hu, Y., Li, J., Bao, J. (2019). Infection of African swine fever in wild boar, China, 2018. Transboundary and emerging diseases 66, 1395-1398.

Lombard, M., Fussel, A. (2007). Antigen and vaccine banks: technical requirements and the role of the European antigen bank in emergency foot and mouth disease vaccination. Revue scientifique et technique-Office international des épizooties 26, 117.

Luby, S.P., Hossain, M.J., Gurley, E.S., Ahmed, B.-N., Banu, S., Khan, S.U., Homaira, N., Rota, P.A., Rollin, P.E., Comer, J.A. (2009). Recurrent zoonotic transmission of Nipah virus into humans, Bangladesh, 2001–2007. Emerging infectious diseases 15, 1229.

Manso-Ribeiro, J., Nunes-Petisca, J., Lopez-Frazao, F., Sobral, M. (1963). Vaccination against ASF. Bull Off Int Epizoot 60, 921-937.

Mariner, J.C., House, J.A., Mebus, C.A., Sollod, A.E., Chibeu, D., Jones, B.A., Roeder, P.L., Admassu, B., van't Klooster, G.G. (2012). Rinderpest eradication: appropriate technology and social innovations. Science 337, 1309-1312.

Mason, P., Piccone, M., McKenna, T.S.-C., Chinsangaram, J., Grubman, M. (1997). Evaluation of a live-attenuated foot-and-mouth disease virus as a vaccine candidate. Virology 227, 96-102.

McOrist, S., Khampee, K., Guo, A. (2011). Modern pig farming in the People's Republic of China: growth and veterinary challenges. Revue Scientifique et Technique-OIE 30, 961.

McVicar, J., 1984. Quantitative aspects of the transmission of African swine fever. American Journal of Veterinary Research 45, 1535-1541.

Mebus, C., Dardiri, A. (1980). Western hemisphere isolates of African swine fever virus: asymptomatic carriers and resistance to challenge inoculation. American journal of veterinary research 41, 1867-1869.

Mebus, C.A. (1988). African swine fever, In: Advances in virus research. Elsevier, 251-269.

Meeusen, E.N., Walker, J., Peters, A., Pastoret, P.-P., Jungersen, G. (2007). Current status of veterinary vaccines. Clinical microbiology reviews 20, 489-510.

Montgomery, R.E. (1921). On a form of swine fever occurring in British East Africa (Kenya Colony). Journal of comparative pathology therapeutics 34, 159-191.

Moutou, F. (2014). The second eradication: Rinderpest. Bulletin de la Société de pathologie exotique 107, 137-138.

Mulumba-Mfumu, L., Goatley, L., Saegerman, C., Takamatsu, H.H., Dixon, L. (2016). Immunization of African Indigenous Pigs with Attenuated Genotype I African Swine Fever Virus OURT 88/3 Induces Protection Against Challenge with Virulent Strains of Genotype I. Transboundary emerging diseases 63, e323-e327.

Mur, L., Gallardo, C., Soler, A., Zimmermman, J., Pelayo, V., Nieto, R., Sánchez-Vizcaíno, J.M., Arias, M. (2013). Potential use of oral fluid samples for serological diagnosis of African swine fever. Veterinary microbiology 165, 135-139.

Nampanya, S., Khounsy, S., Abila, R., Young, J., Bush, R., Windsor, P. (2016). Financial impacts of foot-and-mouth disease at village and national levels in Lao PDR. Transboundary and Emerging Diseases 63, e403-e411.

Neilan, J., Zsak, L., Lu, Z., Kutish, G., Afonso, C., Rock, D.L. (2002). Novel swine virulence determinant in the left variable region of the African swine fever virus genome. virology 76, 3095-3104.

Neilan, J.G., Zsak, L., Lu, Z., Burrage, T.G., Kutish, G.F., Rock, D.L. (2004). Neutralizing antibodies to African swine fever virus proteins p30, p54, and p72 are not sufficient for antibody-mediated protection. virology 319, 337-342.

Netherton, C.L., Goatley, L.C., Reis, A.L., Portugal, R., Nash, R.H., Morgan, S.B., Gault, L., Nieto, R., Norlin, V., Gallardo, C. (2019). Identification and immunogenicity of African swine fever virus antigens. Frontiers in immunology 10, 1318.

Njeumi, F., Taylor, W., Diallo, A., Miyagishima, K., Pastoret, P.-P., Vallat, B., Traore, M. (2012). The long journey: a brief review of the eradication of rinderpest. Rev Sci Tech 31, 729-724.

Nor, M.M., Gan, C., Ong, B. (2000). Nipah virus infection of pigs in peninsular Malaysia. Revue Scientifique et Technique-Office International des Epizooties 19, 160-165.

Normile, D. (2008). Driven to extinction (Science, American Association for the Advancement of Science).

Normile, D. (2018). Arrival of deadly pig disease could spell disaster for China (Science, American Association for the Advancement of Science).

O'Donnell, V., Holinka, L.G., Gladue, D.P., Sanford, B., Krug, P.W., Lu, X., Arzt, J., Reese, B., Carrillo, C., Risatti, G.R. (2015a). African swine fever virus Georgia isolate harboring deletions of MGF360 and MGF505 genes is attenuated in swine and confers protection against challenge with virulent parental virus. Virology 89, 6048-6056.

O'Donnell, V., Holinka, L.G., Krug, P.W., Gladue, D.P., Carlson, J., Sanford, B., Alfano, M., Kramer, E., Lu, Z., Arzt, J. (2015b). African swine fever virus Georgia 2007 with a deletion of virulence-associated gene 9GL (B119L), when administered at low doses, leads to virus attenuation in swine and induces an effective protection against homologous challenge. Virology 89, 8556-8566.

O'Donnell, V., Holinka, L.G., Sanford, B., Krug, P.W., Carlson, J., Pacheco, J.M., Reese, B., Risatti, G.R., Gladue, D.P., Borca, M.V. (2016). African swine fever virus

Georgia isolate harboring deletions of 9GL and MGF360/505 genes is highly attenuated in swine but does not confer protection against parental virus challenge. Virus Research 221, 8-14.

OIE (2013). Manual of Diagnostic Tests and Vaccines for Terrestrial Animals. Paris: OIE. 2013. In Foot and Mouth Disease.

Otte, M., Nugent, R., McLeod, A. (2004). Transboundary animal diseases: Assessment of socio-economic impacts and institutional responses. Food and Agriculture Organization (FAO), 119-126.

Oura, C., Denyer, M., Takamatsu, H., Parkhouse, R. (2005). In vivo depletion of CD8+ T lymphocytes abrogates protective immunity to African swine fever virus. General Virology 86, 2445-2450.

Pankhurst, R., 1966. The great Ethiopian famine of 1888–1892: a new assessment. Journal of the History of Medicine and Allied Sciences 21, 95-124.

Pendell, D.L., Marsh, T.L., Coble, K.H., Lusk, J.L., Szmania, S.C. (2015). Economic assessment of FMDv releases from the national bio and agro defense facility. PLoS One 10.

Pickering, B.S., Hardham, J.M., Smith, G., Weingartl, E.T., Dominowski, P.J., Foss, D.L., Mwangi, D., Broder, C.C., Roth, J.A., Weingartl, H.M. (2016). Protection against henipaviruses in swine requires both, cell-mediated and humoral immune response. Vaccine 34, 4777-4786.

Plowright, W., Ferris, R. (1962). Studies with rinderpest virus in tissue culture: the use of attenuated culture virus as a vaccine for cattle. Research in Veterinary Science 3, 172-182.

Quembo, C.J., Jori, F., Vosloo, W., Heath, L. (2018). Genetic characterization of African swine fever virus isolates from soft ticks at the wildlife/domestic interface in Mozambique and identification of a novel genotype. Transboundary emerging diseases 65, 420-431.

Rahman, M.A., Hossain, M.J., Sultana, S., Homaira, N., Khan, S.U., Rahman, M., Gurley, E.S., Rollin, P.E., Lo, M.K., Comer, J.A. (2012). Date palm sap linked to Nipah virus outbreak in Bangladesh, 2008. Vector-Borne and Zoonotic Diseases 12, 65-72.

Reis, A.L., Abrams, C.C., Goatley, L.C., Netherton, C., Chapman, D.G., Sanchez-Cordon, P., Dixon, L.K. (2016). Deletion of African swine fever virus interferon inhibitors from the genome of a virulent isolate reduces virulence in domestic pigs and induces a protective response. Vaccine 34, 4698-4705.

Robinson, L., Knight-Jones, T.J.D., Charleston, B., Rodriguez, L.L., Gay, C.G., Sumption, K.J., Vosloo, W. (2016). Global Foot-and-Mouth Disease Research Update and Gap Analysis: 3 - Vaccines. Transboundary and Emerging Diseases 63, 30-41.

Rock, D.L. (2017). Challenges for African swine fever vaccine development. Veterinary Microbiology 206, 52-58.

Rodríguez, J.M., Yáñez, R.J., Almazán, F., Viñuela, E., Rodríguez, J.F. (1993). African swine fever virus encodes a CD2 homolog responsible for the adhesion of erythrocytes to infected cells. Virology 67, 5312-5320.

Rodriguez, L.L., Grubman, M.J. (2009). Foot and mouth disease virus vaccines. Vaccine 27, D90-D94.

Rossiter, P.B., Al Hammadi, N. (2009). Living with transboundary animal diseases (TADs). Tropical animal health and production 41, 999.

Roth, J.A. (2011). Veterinary vaccines and their importance to animal health and public health. Procedia in Vaccinology 5, 127-136.

Roth, J.R., Galyon, J (2019). Emerging Diseases of Food Animals Threaten Global Food Security. In International Affairs Forum: Food and Water Security (Center for International Relations).

Rowlands, R.J., Michaud, V., Heath, L., Hutchings, G., Oura, C., Vosloo, W., Dwarka, R., Onashvili, T., Albina, E., Dixon, L.K. (2008). African swine fever virus isolate, Georgia, 2007. Emerging infectious diseases 14, 1870.

Rweyemamu, M., Roeder, P., Mackay, D., Sumption, K., Brownlie, J., Leforban, Y., Valarcher, J.F., Knowles, N., Saraiva, V. (2008). Epidemiological patterns of foot–and–mouth disease worldwide. Transboundary and emerging diseases 55, 57-72.

Sánchez-Vizcaíno, J.M., Mur, L., Martínez-López, B. (2013). African swine fever (ASF): five years around Europe. Veterinary Microbiology 165, 45-50.

Sifferlin, A. (2017). This new group wants to stop pandemics before they start. Time. Retrieved June 8, 2020, from https://time.com/4639245/cepi-ebola-zika-epidemic/.

Speer, N. (2019). The impact of African swine fever goes far beyond just the disease. Beef magazine. Retrieved June 8, 2020, from https://www.beefmagazine.com/outlook/china-pork-production-just-how-bad-it.

Stenfeldt, C., Segundo, D.S., de los Santos, T., Rodriguez, L.L., Arzt, J. (2016). The pathogenesis of foot-and-mouth disease in pigs. Frontiers in veterinary science 3, 41.

Stone, S., Hess, W. (1967). Antibody response to inactivated preparations of African swine fever virus in pigs. American journal of veterinary research 28, 475.

Thomas, L.F., Bishop, R.P., Onzere, C., Mcintosh, M.T., Lemire, K.A., de Glanville, W.A., Cook, E.A.J., Fèvre, E.M. (2016). Evidence for the presence of African swine fever virus in an endemic region of Western Kenya in the absence of any reported outbreak. BMC veterinary research 12, 192.

Thompson, D., Muriel, P., Russell, D., Osborne, P., Bromley, A., Rowland, M., Creigh-Tyte, S., Brown, C. (2002). Economic costs of the foot and mouth disease outbreak in the United Kingdom in 2001. Revue scientifique et technique-Office international des epizooties 21, 675-685.

Thomson, G., Vosloo, W., Bastos, A. (2003). Foot and mouth disease in wildlife. Virus research 91, 145-161.

Vergne, T., Chen-Fu, C., Li, S., Cappelle, J., Edwards, J., Martin, V., Pfeiffer, D.U., Fusheng, G., Roger, F.L. (2017). Pig empire under infectious threat: risk of African swine fever introduction into the People's Republic of China. Veterinary Record.

Vilanova, E., Tovar, A.M., Mourão, P.A. (2019). Imminent risk of a global shortage of heparin caused by the African Swine Fever afflicting the Chinese pig herd. Journal of Thrombosis and Haemostasis 17, 254-256.

Villeda, C., Gómez-Villamandos, J., Williams, S., Hervas, J., Wilkinson, P., Vinuela, E. (1995). The role of fibrinolysis in the pathogenesis of the haemorrhagic syndrome produced by virulent isolates of African swine fever virus. Thrombosis haemostasis 73, 112-117.

Villeda, C., Williams, S., Wilkinson, P., Vinuela, E. (1993). Consumption coagulopathy associated with shock in acute African swine fever. Archives of virology 133, 467-475.

Wang, T., Sun, Y., Qiu, H.J. (2018). African swine fever: an unprecedented disaster and challenge to China. Infectious diseases of poverty 7, 111.

Weingartl, H.M., Berhane, Y., Caswell, J.L., Loosmore, S., Audonnet, J.-C., Roth, J.A., Czub, M., (2006). Recombinant nipah virus vaccines protect pigs against challenge. Journal of virology 80, 7929-7938.

Wilkinson, P., Wardley, R., Williams, S. (1983). Studies in pigs infected with African swine fever virus (Malta/78). In: CEC/FAO Expert Consultation on African Swine Fever Research, 74-84.

Zhao, D., Liu, R., Zhang, X., Li, F., Wang, J., Zhang, J., Liu, X., Wang, L., Zhang, J., Wu, X. (2019). Replication and virulence in pigs of the first African swine fever virus isolated in China. Emerging microbes & infections 8, 438-447.

Zhou, X., Li, N., Luo, Y., Liu, Y., Miao, F., Chen, T., Zhang, S., Cao, P., Li, X., Tian, K. (2018). Emergence of African swine fever in China, 2018. Transboundary and emerging diseases 65, 1482-1484.

Chapter 6

Neonatal Vaccination and Maternal Immunization

Heather L. Wilson and Volker Gerdts*

Vaccine and Infectious Disease Organization, International Vaccine Centre, University of Saskatchewan, 120 Veterinary Road, Saskatoon, SK, Canada

*Corresponding author: volker.gerdts@usask.ca

DOI: https://doi.org/10.21775/9781913652593.06

Abstract

Neonates of all species are highly susceptible to infectious diseases. Current evidence suggests that neonates of most species are fully capable of mounting effective immune responses. Indeed, one of the reasons why neonates greatly suffer from infectious disease might be the fact that immune responses in this age group are often too strong contributing to immune pathology and non-physiological responses rather than simply clearing the infection. This hypothesis suggests that neonates may effectively respond to vaccination when properly stimulated, and that response to vaccination in neonates can be, and should be, driven through the use of adjuvants and other innate immune modulators that help guide the response towards a protective and long-lasting response. Here, we review some of the current strategies to enhance the effectiveness of vaccines in neonatal animals in the presence of passive maternal immunity. Colostral and lactogenic passive immunity are particularly important mechanisms to protect against enteric diseases in the very young, in particular through the transfer of maternal antibodies. Immunizing the mother to increase the passive transfer has been utilized for centuries to enhance the level and duration of protection in the neonate. Here, we will review some innovative strategies for maternal and neonatal vaccines against common infectious diseases.

The neonatal immune system

Neonates of most species are born with the ability to mount effective immune responses *in utero* and in early life. Generally, mammals with a longer period of gestation have a more mature immune system at birth than do species with short gestational periods (Banks 1981). Especially in our large animal species, third trimester fetuses are capable of distinguishing self from non-self and mounting effective immune responses to *in utero* infections and vaccines. Examples of such infections include prominent viral infections such as Bovine Virus Diarrhea Virus (BVDV), Equine Herpesvirus (EHV), or Porcine Reproductive and Respiratory Syndrome Virus (PRRSV), bacterial infections such as brucellosis or leptospirosis, and parasitic infections such as toxoplasmosis and *neospora caninum* (2). In most species, the last trimester fetus effectively responds to and often is able to clear the infection, demonstrating that the third-term fetus is capable of mounting antigen-specific immune responses. For example, studies in horses shows that neutralizing antibodies are produced by the equine fetus when antigens are injected after 200 days of gestation (3). This approach has been industrialized in the poultry industry, where chicken embryos are being immunized *in ovo* for diseases such as Marek 's disease (4), Newcastle disease (5), Infectious Bursal Disease (6), and coccidiosis (7), with minimal effects on hatchabilty. In livestock species, the concept of immunizing the fetus has been largely underutilized, despite the fact that it might provide a potential strategy to prevent infections that are associated with labour and the early neonatal period. Sporadic observations in horses, pigs, and cattle have demonstrated that last trimester fetuses can respond to immunization (3). To test this concept with a novel DNA vaccine and a subunit vaccine, an experimental DNA vaccine was given orally to fetal lambs in the last trimester. The experimental vaccine was highly effective in inducing both humoral and cell-mediated immune response in these fetuses prior to birth and protected immunized animals from viral challenge with bovine herpesvirus 1 (8). The vaccine was administered orally into the amniotic fluid and immune responses were found to carry over into the neonatal period, with anamnestic responses detectable after another boost at one month of age (9). Thus, these studies confirm that fetuses of our domestic livestock species are equipped to respond to either infection or immunization with an innate and adaptive immune response.

At birth, the neonatal immune system faces a number of challenges. These including i) exposure to large amounts of foreign antigen in the microbiota and environment, ii) high levels of cortisol in preparation for labour, as well as the need to iii) quickly establish immunological tolerance by closing the gut and establishing pathways to sample free antigen via secondary lymphoid organs such as tonsils and Peyer's patches. While the neonatal immune system has often been referred to as "immature" or not "fully competent", recent observations suggest that the neonatal immune system is well equipped to respond to infection. However, neonatal and adult response can differ significantly, and in particular the type and magnitude of the response often differ from what one can find in adult animals, with these responses sometimes being disproportionally high or in other cases even non-physiological and harmful which can lead to disease manifestation (10) .

Innate and adaptive immune responses in the neonate

The state of the neonatal immune system is influenced by a wide range of genetic and epigenetic factors such as maternal cytokines and chemokines, microbiota, host genetics, co-infections and maturation of immune cells (Table 1). However, this varies greatly amongst our domestic species, and only little is known about the state of the innate immune response for each species. A wide variety of factors including colostral and lactogenic immunity (maternal antibodies), growth factors, innate lymphoid cells, T cells and cytokines play a role in setting the innate immune response. For example, pregnancy associated effectors such as interferon tau, progesterone, pregnancy associated glycoproteins and chaperonins regulate the immune response both locally in the uterus, as well as systemically in peripheral immune cells and other tissues (11). Neutrophil function is subdued and the frequency of neutrophils is lower in neonates compared to adults for most species. Interestingly, neonatal neutrophils are not able to form neutrophils NETs (reviewed in (12)) and display decreased expression of adhesion molecules such as CD62L, L-selectin and MAC-1. Functions such as phagocytosis, neutrophilic burst and diminished capacity of degrading intracellular infective agents are all features of neonatal neutrophils (reviewed in (12)). Neonatal antigen presenting cells express lower levels of MHC and co-stimulatory molecules and show a tendency towards Th2 immune responses (reviewed in (13) and (14)). In addition, a higher ratio of plasmacytoid versus conventional dendritic cells (DCs) can be found in neonates, combined with lower chemotaxis and reduced extravasation (12). The complement

Table 1. Examples of immunological features of the neonatal immune system

Immunity	Effector cell	Mechanisms
Innate	Complement system	Lower concentrations of some complement proteins
	Neutrophils	Lower neutrophil frequency; reduced neutrophil function including phagocytosis, neutrophilic burst; phagolysis; no functional neutrophil nets; decreased expression of adhesion molecules
	Antigen-presenting cells	Lower levels of MHC and co-stimulatory molecules; higher ratio of plasmacytoid versus conventional dendritic cells, reduced chemotaxis, reduced extravasation
Adaptive	B cells/antibody-secreting cells	Delayed follicular dendritic cell maturation; limited development and expansion of follicular T helper cells; delayed development of germinal centre B cells and plasma cells; limited antibody-repertoire at birth; altered expression of maturation markers
	T cells	Low Th1:Th2 ratio; high FoxP3 to Th17 ratio; high levels of IL-10 and TGF-b

system in neonates is less functional due to lower concentrations of complement proteins C1q, C3, C4, and factor B. Natural killer cells are found at high frequencies in neonates, however, their responsiveness is reduced and NK frequency often takes a long time before it reach adult levels. Other innate lymphoid cells (ILCs), such as ILC2, play an important role along the mucosa during the establishment of the microbiome in human infants (15). Many ILCs and other innate mediators are found in colostrum and milk, and depending on the placenta type are transferred shortly before or after birth. In particular, maternal ILCs are thought to protect the newborn against infections by helping to maintain the integrity of the mucosal surfaces and assist in shaping the microbiome (12,16).

Trained innate immunity is the result of long-term epigenetic changes of chromatin remodeling and functional metabolic reprograming of innate immune cells upon priming that triggers enhanced or suppressed response to secondary stimuli. The priming response is often associated with pathogens or pathogen-associated molecular patterns (PAMPs), but there is evidence that food components, possibly acting as PAMPS, can also induce trained immunity. For instance, bovine milk- and bovine immunoglobulin-stimulation of human monocytes, that were then rested for 5 days and restimulated with TLR ligands, showed induced trained immunity as measured by induced TNF-alpha and IL-6 production (17). Limited studies are available that indicate whether trained immunity promotes the health of the newborn although studies in young animals are being increasingly investigated. Guerra-Maupome immunized six to eight-week old Holstein calves with *Mycobacterium bovis* BCG strain Danish via aerosol inoculation (18). One month later, peripheral blood mononuclear cells (PBMCs) were isolated and re-stimulated with LPS or Pam3CKS4. The results confirmed that *in vivo* BCG exposure significantly enhanced expression of the monocyte-derived pro-inflammatory cytokine production in response to *ex vivo* stimulation with LPS and Pam3CSK4 (18). The metabolic programing of the PBMCs were also affected such that they had elevated glucose consumption and lactate production relative to PBMCs from unvaccinated animals suggesting induction of trained immunity. Beta-glucans from yeast can induce trained immunity. In another study, Angulo et al. stimulated newborn goats on 1 and 4 days of age with 50 mg/kg β-glucan from *Debaryomyces hansenii* (β-Dh). The goats were then i.p. challenged on day 7 with LPS and they showed increased respiratory burst, increased plasma IL-1β, IL-6, and TNFα, and increased transcription of the macrophage surface markers in respiratory burst activity (19). These results and others suggest that newborns are responsive to trained immunity and therefore its use should be investigated as a potential means to limit livestock disease without the use of antibiotics (19,20). The development of targeted approaches to engage innate training in food animals, with the practical goal of enhancing the capacity to limit disease without the use of antibiotics, is an area which deserves continued attention.

Strong adaptive immune responses can be found in all neonates of our domestic livestock species. However, the induction of primary antibody responses in neonates in general is hindered by delayed follicular dendritic cell maturation, limited development and expansion of T follicular helper cells (Tfh), and delayed development of germinal centre B cells and plasma cells (reviewed in (21); (22); (23); (24)). T cell immunity in the neonate is characterized by a low Th1:Th2 ratio, and high FoxP3 to Th17 ratios. Neonatal IL-10 and TGFβ levels are significantly elevated compared to adults. However, neonates can mount strong Th1 immune responses when properly stimulated. Furthermore, there is a great variation amongst our domestic livestock species and many of the above observations were made in mice and humans, and thus can only be extrapolated to large animals or poultry. Evidence in pigs, however, suggests that the antibody repertoire develops early in life in Peyer's patches in the gut, and in particular SIgA are provided in large quantities early in life by intestinal Peyer's patches (25). In adult cattle, 90% of circulating B cells express CD21 (an activating receptor) and CD32 (an inhibitory receptor) whereas the number of circulating B cells in calves was shown to be less than 20% of that found in adults (26). Further, newborn calves expressed lower CD21 and CD32 receptors which took 6 weeks and 3 weeks, respectively, to reach adult expression levels. The altered CD32 to CD21 ratios in the young animal B cells versus the adult animals B cells may impact their humoral immune response capacity (26). The frequency of circulating B cells in pigs does not appear to vary with age (27).

Maternal and passive immunization

Strategies to induce colostral and lactogenic immunity

Passive immunity is achieved by trafficking of mucosal lymphocytes to the mammary gland, which provides cells, antibodies, and innate immune molecules to be transferred to the suckling offspring. The immature intestinal epithelium of the newborn cattle, pigs, horses, and sheep allows the non-selective transfer of IgG and macromolecules across the intestinal lumen into the lymphatics generally up to 24 hours after birth (28-30). These animals are born agammaglobulinemic and reliant upon passive transfer of colostral and lactogenic immunoglobulins for protection against disease (31). One report calculated that calves must consume enough colostrum to achieve at least 5.7 g/dl serum IgG to be protected from bovine respiratory disease before 5 weeks of age (32). Calves that fail to quickly achieve elevated serum IgG suffer from 'Failure of Passive Transfer' (FPT) (33). Delaying colostrum feeding within 12 h of life can decrease the passive transfer of IgG and may delay the colonization of bacteria in the intestine leaving the pre-weaning calf vulnerable to infections (34). Vaccines administered to pregnant cows have been shown to promote protective immunity to suckling calves against a number of pathogens including rotavirus (35) and colibacillosis (36) with many vaccines being commercially available. Like cows, ewes vaccinated with K99 pili from enteropathogenic *Escherichia coli* (EPEC) or live attenuated *Salmonella* have demonstrated passive protection in suckling lambs (37,38). In dogs, immune transfer during pregnancy from dam to the fetus is restricted due to the endotheliochorial placentation such that less than 7% of the final serum immunoglobulin G (IgG) concentration at 24–48 hr of age is acquired in prior to birth (39). Maternal passive immunity has been shown to be effective in protecting puppies against canine parvovirus (40). For many of these studies, it has not been directly ascertained whether antibodies and antibody-secreting cells are transferred to the offspring via colostrum or milk although it is assumed in the colostrum as gut-closure tends to occurs within 1-2 days after birth (41).

Passive transfer of maternal immunity also occurs in non-mammal animals. In chickens, maternal immunoglobulins are transferred to the pre-hatched chick through the yolk sac (42) such that 3 days after hatch, the circulating IgY in the chick is approximately 1/3rd that of the chicken (43). Studies have shown that vaccinating the adult specific pathogen-free broiler with Salmonella enterica serovar Enteritidis ghost vaccines led to significant protection of 7-day old chicks against infectious S. Enteritidis. Others have shown that vaccinating broiler breeders with a bivalent live and an inactivated fowl adenovirus vaccine protected chicks against developing inclusion body hepatitis upon challenge with a lethal dose of fowl adenovirus serotypes 2, 8b, 11 through passive transfer via the eggs (44). Likewise, vaccinated fish can transfer passive immunity in the yolk that is sufficient to protect their young against disease (45). Further, antibodies may be transferred to fry from skin (46) and from the mouth cavity (47). Teleost fry that underwent mouth brooding from a mother previously exposed to *Ichthyophthirius multifiliis* showed increased survival rates after being exposed to infective tomites of *I. multifiliis* relative to fry that did not undergo mouth brooding (47). Thus, passive transfer can protect the young of many species against infectious disease.

The stage of gestation and the timing of infection by a pathogen or vaccination to promote maximal lymphocyte trafficking to the mammary gland is still being defined for each species. One study showed that gilts that were orally administered infectious PEDV in the 2nd trimester had significantly higher levels of circulating anti-PEDV IgA and IgG antibodies, antibody-secreting cells, and PEDV virus neutralizing (VN) antibodies relative to gilts infected during the 1st trimester, 3rd trimester or mock-infected gilts (48). Further, 100% of piglets born from gilts infected during the 2nd trimester survived PEDV challenge relative to 87.2, 55.9, and 5.7% survival for challenged pigs born from gilts that were infected during the 1st or 3rd trimester or mock-infected, respectively. Piglet survival positively correlated with PEDV IgA antibodies, antibody-secreting cells and VN antibodies in milk and PEDV IgA and IgG antibodies in piglet serum (48). The cell frequencies, immune parameters of innate immune cells, and serum cytokine profiles were shown to be different throughout pregnancy (48). Another study showed that daily oral vitamin A supplementation of PEDV-infected gilts in the 3rd trimester enhanced serum and milk IgA in the dams and lactogenic immune protection of suckling offspring (49). Others have shown that sows immunized orally with PEDV-coated microspheres at the beginning of the 3rd trimester followed by a booster immunization 2 weeks prior to farrowing with intramuscular killed PEDV vaccines had significantly higher neutralizing antibody titers milk than did control sows and that 90% of their suckling piglets were protected from challenged with infectious PEDV (50). Likewise, early or late gestation vaccination of cows has also been investigated. One study showed that cows vaccinated in the first 3 months (early gestation vaccination (EGV)) versus cows vaccinated in at 6.5 months (late gestation vaccination (LGV)) had disparate transfer of passive immunity to suckling calves (51). The average neutralizing antibody titers did not differ between the EGV and LGV groups regardless of whether booster vaccinations were included but calves that suckled from LGV cows had significantly higher mean maternally derived antibody titers and they had longer duration than that of the EGV calves and this could potentially provide offspring greater protection of the offspring. Others showed that sows immunized 40 days prior to farrowing at the Houhai acupoint (between the tail and the anus) with PDCoV-inactivated vaccine exihibited 87% protective efficacy in suckling piglets in response to PDCoV infectious challenge (52). Understanding how the timing of viral infection and/or vaccination during pregnancy impacts maternal immunity will allow researchers to pinpoint when the dam is the most immunologically responsive to vaccines to protect her and her offspring via colostral or lactogenic immunity.

Maternal cells are present in the colostrum and milk, and they may play a role in protecting suckling offspring. A report in calves showed that labeled maternal colostral cells were detected in the epithelium, domes and sinuses around jejunal Peyer's patches but they were only detected in the ileal Peyer's patch in the sinuses suggesting region-specific uptake (53). Studies in pigs showed that colostral lymphoid cells from the dam (but not colostral cells from others sows) were absorbed by the piglet digestive tract and migrated to the mesenteric lymph nodes (54) and that maternal T lymphocytes were detectable in piglet serum after suckling (55). These lymphocytes express activation markers (56) and studies showed that calves that received whole colostrum had fewer CD11a positive lymphocytes in circulation and a higher percentage of lymphocytes expressing the activation markers CD25 and CD26 after birth relative to calves fed cell-free colostrum (57). Further, calves fed colostrum with maternal cells had mixed-lymphocyte reaction (MLR) after 24 hour but calves fed cell-free colostrum took 2-3 weeks took to show a MLR response suggesting that maternal colostral cells stimulated development of the neonatal immune system (58). A recent report showed that innate lymphoid cells (ILCs) have been identified in human colostrum (59). Whether they are present in colostrum from livestock species and whether they contribute to the innate immune response in the neonate must still be determined. More studies across livestock and domestic animal species are required to determine the extent to which maternal cells from colostrum or milk contribute to cell-mediated immunity and induction of protective immunity in the suckling offspring (33).

Non-immunoglobulin and non-cellular components in colostrum and milk may contribute to passive immunity. A wide-range of growth factors, cytokines and antimicrobial proteins and peptides, such as lactoferrin, defensins and cathelicidins, IGF-I and -II, leptin, alpha-lactalbumin, beta-lactoglobulin, lactoperoxidase, lysozyme, and complement 3, and many others, are present in colostrum and/or milk that may promote or otherwise regulate offspring performance (60-65). However, there is a paucity of data establishing whether these macromolecules directly contribute to passive immune protection in livestock and domestic species. More studies are needed to determine whether further supplementation of these select macromolecules can promote lactogenic immunity.

Promotion of passive immunity through supplementation

Supplementing offspring with antibodies and macromolecules via colostrum, spray-dried plasma or egg yolks can be an important form of externally sourced hyperimmune products to control infectious disease in offspring (66,67). For example, providing calves with a supplementary colostrum meal in addition to nursing was shown to be effective in maximizing passive immunity transfer and reducing calf disease incidence and mortality in early life (68). Of course, bovine colostrum (replacer or spray-dried supplements) or egg yolks will only contain antibodies against pathogens to which the cow and hen have been exposed and therefore an industry is developing to obtain hyperimmune products from immunized animals for the purpose of passive immunization of neonates against select targets. Research has shown that bovine colostrum or egg yolks from immunized cows and birds can reduce enterotoxigenic *Escherichia coli* (ETEC) infection (69), rotavirus-induced diarrhea (70) and coronavirus-induced diarrhea (71) when fed to calves and pigs. Studies in puppies have shown that obtaining and using hyperimmune sera from canine colostrum or serum (or plasma) banks, bovine colostrum or hyperimmune egg powder can positively impacted growth, reduced morbidity in the neonatal period and decreased risk of canine parvovirus infection (72). Passive immune supplementation is also critically important in farmed species that lack an adaptive immune. For example, shrimp are susceptible to mortality due to infection by *Vibrio* spp. To combat this, the industry is researching feeding adult shrimp egg yolk powders with anti-*Vibrio* hyperimmune antibodies to increase the survival of the young (73). Of course, producers must weigh whether the benefits of passive immunity outweigh the costs to obtain and administer them. However, with industry-specific challenges such as the necessity for artificial rearing and formula feeding due to increasingly large sow litter and large-scale fish farming, etc., more research will certainly be needed to provide more extensive passive immunity and feed for newborns.

There is evidence that supplementation of the sow's diet can impact piglet passive immunity. Sows supplemented with short-chain short-chain fructo-oligosaccharide (scFOS) significantly transferred acetate, propionate, calerate and caproate to suckling offspring (74). Suckling piglet's experienced increased caecal mucin-secreting goblet cells, increased ileal IFNg production and improved body composition (74). When immunized via the oral route with live-attenuated *L. intracellularis* vaccine at weaning, piglets from sows supplemented with scFOS had elevated *L. intracellularis*-specific serum IgA but not IgG. Together these results suggest that maternal diet composition can impact not only passive immunity but also the active immune response in the weanling piglet (74) and the impact on other species should also be investigated.

Neonatal immunization

A delicate balance must take place at birth. Commensal flora must populate the intestinal tract, nasal and vaginal cavities, skin, etc. and the neonatal immune system must facilitate this microbial colonization while avoiding unnecessary and potentially hazardous pro-inflammatory responses. At the same time, the newborn is vulnerable to infection from pathogenic organisms. To compensate, the neonate receives passive antibodies in maternal colostrum which protects the neonates from diseases to which the dam has been exposed. Despite this passive protection, it is estimated that as much as 5% of foals, 5%–10% of calves, and 10%–15% piglets die before weaning (75) with infectious diseases being amongst the leading causes of death. With these incidences, it is not difficult to understand that additional protective measures such as inducing an immune response in the newborn could also help protect against losses against infectious disease.

Challenges into the vaccination of neonates include vertical transmission of pathogens from dam to offspring causing early infection, unpredictability of maternal antibody titres, and excessive skewing towards Th2 immunity which must be overcome (possibly with adjuvants in vaccine formulations). Also, very early exposure to a small dose of antigen may promote tolerance and reduce subsequent immune responses (76) so studies on antigenic dose, timing of exposure and use of vaccine adjuvant are needed to promote immunity and avoid immune tolerance. Further, maternally-derived antibodies (MDA) can interfere with development of an effective neonatal immune response. Interference by MDA has been associated with masking of immunodominant epitopes, regulation of B cell activation and germinal center maturation, and B-cell inhibition through FcγRIIB cross-linking (77,78). Vaccination of calves in the face of MDA can be compromised due to maternally derived immunity interfering with vaccination. However, evidence suggest that effective T- and B-cell memory responses can be primed suggesting that memory T- and B- cells are much less sensitive to MDA (reviewed in (79); (80-82). An interesting concept has been proposed that vaccinating dam and offspring with distinct, non-overlapping epitopes of the same antigen, referred to as pair-epitope vaccination, may avoid MDA and allow offspring to develop their own protective immune response (83). So far, no vaccines using this approach have been studied but it appears to be a reasonable approach worthy of study.

Examples of innovative neonatal vaccines

Vaccination strategies for human infants have been recently reviewed in an excellent review by Kollmann et al. (13). Several examples of live vaccines in humans have demonstrated broadly enhanced host resilience against infection beyond their specific pathogen target (84,85). Most likely these vaccines induce cross-reactive T cells and antibodies (e.g. heterologous immunity) and activation of innate immune components (e.g. trained immunity). Strategies to overcome maternal suppression of the neonatal immune response may include using a higher or repeated doses and alternative routes of administration and use of novel adjuvants or specific to promote immunity. Similarly, many new vaccine strategies are currently being evaluated for use in neonatal animals. These include a variety of approaches including live-attenuated, subunit vaccines as well as novel adjuvant platforms that that are specifically designed to induce early-life immunity. Examples of such approaches are listed below:

Modified live vaccines

Modified live vaccines (MLV) are effective vaccines because they replicate within the host and may not require exogenous adjuvants to stimulate the appropriate innate and adaptive immune responses. While they have a low risk of reversion to virulence, a short half-life and require a cold chain to remain effective, they work well, especially if administered intranasally and thus many commercial vaccines using live attenuated pathogens exist for newborns. For example, ONCE PMH® IN (Merck Animal Health) is used to prevent *Mannheimia Haemolytica* and *Pasteurella Multocida* in calves. Further, a protection study showed that a single intranasal administration of a BRSV vaccine INFORCE 3 (Pfizer Animal Health) administered to 3-11 day old Holstein calves could induce protective immunity in the presence of maternal antibodies if the challenge was administered 9 weeks, but not 16 weeks, post vaccination (86). These experiments suggest that mucosal administration of a modified-live virus (MLV) can overcome maternal interference and protect against infection if the period between vaccination and challenge is sufficiently short (87). Similarly, bivalent H1N2 and H3N2 NS1-truncated live-attenuated influenza vaccines have shown promising results in pigs. When newborn piglets were vaccinated by the intranasal route with the truncated NS1 live attenuated influenza A vaccine, they showed significantly reduced viral replication and shedding after challenge with either H1N2 or H3N2, even in the presence of MDA (88). Others showed that a MLV vaccine consisting of BHV1, BVDV, BVDV2, PI3 and BRSV antigens administered intranasally to newborn calves induced mucosal IgA antibodies despite MDA (89). Other examples include canine distemper virus, Newcastle virus, PRRSV and many others. Intramuscular MLV vaccines were efficacious in piglets from herds where PRRSV-1 or -2 were endemic were immunized i.m. with Fostera™ PRRS (Zoetis) at 1 day of age then they were infected with PRRSV-1 and -2 at 70 days of age.

Neutralizing serum antibody titres were not significantly different between vaccinated and unvaccinated groups but anti-PRRSV serum antibody titres and IFNγ secreting cells were significantly elevated in the vaccinated vs unvaccinated groups (90). These results demonstrate that newborns have sufficiently mature immune systems to elicit humoral and cell-mediated immunity to the MLV vaccines even in the presence of maternal antibodies.

The poultry industry has made considerable strides in establishing that *in ovo* vaccination and/or vaccination in the hatchery can produce an immune response and signs of protection against disease. For example, live herpes virus of turkeys (HVT) Rispens vaccine was administered subcutaneously to specific-pathogen-free 1-day-old chickens. At 8 days of age, the birds were challenged with infectious HVT into the abdomen and the vaccine prevented the α-herpesvirus neoplastic Marek's disease (91). Such MLV vaccines continue to show their efficacy, even in very young animals, which makes them attractive to use to protect livestock against evolving pathogens.

Viral and bacterial vectors

Several vector-based platforms have been developed for livestock and domestic animals. Canarypox virus, herpes virus, and modified vaccinia viruses as vectors have proven highly efficacious in the young from a variety of species including dogs, cats, ferrets, pigs, horses, etc. (92-96) with continued development to prove efficacy in newborns. Canarypox vectored canine distemper virus vaccine administered subcutaneously to puppies triggered elevated serum neutralizing antibodies and protection against disease relative to the unvaccinated pups, even in the presence of MDA (92). One-day old piglets vaccinated via intramuscular (but not intranasal) route with three separate adenovirus coding for either gD, gC or gD showed elevated serum neutralizing antibodies even in the presence of MDA and limited protection for challenge at 16 weeks (97).

Bacteria can also be used as live vaccine or vaccine vectors (95,98). *Bacillus subtilis*, a nonpathogenic Gram-positive commensal bacterium, was modified to express the gene COE that codes for a region of the PEDV Spike protein (*B. subtilis-RC*). Oral administration of newborn piglets with *B. subtilis*-RC significantly promoted piglet intestinal development and it induced anti-PEDV local and systemic humoral and cell-mediated immunity immune responses. *B. subtilis*-RC administration to the newborns could also stimulate high level of PEDV neutralizing antibodies against PEDV infection and should be investigate for protection against disease (99). Likewise, oral administration of newborn piglets with *B. subtilis* expressing PCV2 capsid protein had elevated levels of PCV2-specific IgA in the mucosal tissues of the digestive and respiratory tract, and PCV2-specific IgG in serum, significantly increased IgA-secreting cells and frequency of intestinal intraepithelial lymphocytes, $CD3^+$ T lymphocytes and $CD4^+$ T lymphocytes relative to non-immunized newborns. Further vaccine and challenge trials are needed to determine efficacy of vectored vaccine across multiple species in the presence of MDA.

Subunit vaccines

Subunit vaccines offer the advantage of a high safety profile and a relative ease of manufacturing and they are currently being used in a variety of human and veterinary vaccines. To be effective, subunit vaccines require formulation with effective adjuvant or immunomodulators to properly stimulate T cell responses and long-term duration of immunity. As an example, in ovo injection of a purified recombinant protein from *Eimeria acervulina* (3-1E) alone or in combination with expression plasmids encoding the interleukin IL-2, IL-5, IL-17, or IL-18 genes showed significantly decreased oocyst shedding and serum antibody titres compared with that in non-vaccinated and challenged chickens (100). Another mechanism to improve immunogenicity of subunit or epitope vaccines can be to fuse the peptide/ protein to a PAMP such as TLR2 agonist with proper linkers (101). Vaccine platforms are available that enable rapid manufacturing and expression of subunit vaccines for animals (102). Subunit vaccines using adjuvant platforms would have the advantage of being immediately available in the face of disease outbreaks and by having an established safety record, could be quickly approved by regulators.

Virus-like particles

Virus-like particles (VLPs) are assembled from recombinant proteins and lack a viral genome. The proteins (endogenous to the virus or from foreign pathogens that use the VLP as a vector) represent protective antigens which are often formulated with proper adjuvants to induce or direct effective immune responses. For example, PAMPs present on or within the VLP can act as adjuvants to augment the vaccine response (103). Studies in gnotobiotic piglets showed that a VLP expressing human norovirus and formulated with immunostimulating complexes (ISCOMs) and mutant

E. coli T toxin (mLT) as mucosal adjuvants induced significant humoral and cell-mediated intestinal and systemic immunity and protection against disease compared to controls (104). Similarly, gnotobiotic calves administered bovine VLPs plus mLT intranasally had elevated fecal IgA antibodies and showed partial protection after challenge (105). Studies must continue to establish whether VLPs protect conventionally reared livestock and show protection against veterinary diseases but these results are very encouraging for livestock industries.

Nanoparticle based-vaccines

Nanoparticle based vaccines generated from virus-like particles and biodegradable and biocompatible polymers and liposomes have been shown to promote antigen internalization and processing by antigen presenting cells leading to effective induction of innate and adaptive immunity (106). A polyanhydride nanoparticle encapsulating glycoproteins from BRSV administered to 2-3 week old calves via the intranasal route showed protection against BRSV aerosol challenge (107). Vaccinated calves showed reduced pathology in the lungs, reduced viral burden, and decreased virus shedding compared to unvaccinated control calves suggesting that the nanovaccine was highly immunogenic (107). Others recognized that a nanoparticles consisting of ribonucleoprotein complexes coding for RSV P protein (called a nano-ring) could be tested for efficacy against BRSV which shares a high degree of homology (108). Calves (colostrum free and without anti-BRSV antibodies) were injected via the intramuscular route alone or including the intranasal route with nano-ring formulated with Montanide^{TM} (109). The vaccinated calves developed systemic and intranasal anti-N protein antibodies and cellular immunity in the lymph nodes and moderate protection against BRSV challenge. Thus, nanoparticles are being investigated for use a neonatal vaccines in the presence and absence of MDA.

have being investigated as an evolving technology for the fast development of vaccines against newly emerging pathogens and there is evidence that it triggers immunity in newborn animals. For example, piglets were immunized intradermally with rabies virus encoding a glycoprotein (RABV-G) 3 days after birth with a booster 3 weeks later (110). All immunized piglets rapidly seroconverted to neutralizing antibody titers at day 28 that were comparable to those induced by full doses of the licensed human vaccine (110). This study demonstrates the feasibility of a non-replicating mRNA rabies vaccine in newborn large animals.

DNA vaccination has also shown promise in overcoming MDA. Piglets (5-6 days of age) with high MDA for PRV were immunized via the intramuscular route with DNA plasmids coding for three PRV glycoproteins or not vaccinated (111). The vaccinated piglets showed a low primary serum antibody response but a strong anamnestic humoral response after the booster DNA vaccine relative to the unvaccinated piglets. These data suggest that despite high MDA, newborn piglets administered DNA vaccine show efficient antibody priming and anamnestic in response. Effective nucleic acid vaccines may offer a unique opportunity to design protective vaccines for very young animals even in the presence of MDA.

Adjuvants

An area of significant improvement in the development of neonatal vaccines is the development of novel adjuvants (112). Adjuvants can be used to drive the immune response towards a certain type of immunity, they can increase the onset of immunity, and they can be used to enhance the overall magnitude and extend the duration of the response. However, especially in the context of vaccine for neonates, they are crucial to increase the effectiveness of the vaccine. Combination adjuvants have been developed to drive the immune response in the neonate towards a balanced, Th1 or even Th17 response in a number of animal models and for a number of pathogens. For example, the combination of a biodegradable polymer (PCEP), host defense peptides and Toll-like receptor ligands was shown highly effective in neonatal chicks, mice, pigs, and calves (113-119). Importantly, the adjuvant improved vaccine responses in the neonate even in the presence of maternal antibodies and after mucosal immunization (119,120). Other examples include CpG ODN or other TLR-Ligands that can shift the response towards a Th1 response which can promote vaccine efficacy (121).

Mucosal routes of immunization

Improved targeting the intestine and improved mucosal adjuvants are being investigated as a means to overcome maternal interference and to promote the newborn animal innate and adaptive immune responses. Lambs (<4 days old) immunized directly into surgically isolated intestinal segments (i.e. enteric immunization) with a plasmid encoding a truncated secreted version of bovine herpesvirus-1 gD (tgD) developed antibody and T-cell responses to the tgD glycoprotein (122) despite high levels of passively acquired serum antibodies. This enteric immunization primed a humoral response in the lungs confirming that the MALT in newborn lambs is functional even in the face of MDA. Oral immunization with genetically-modified bacteria has also shown to be effective in protecting newborn piglets against PEDV. Genetically engineered *Lactobacillus casei* expressing a dendritic cell (DC)-targeting peptide fused with region of a porcine epidemic diarrhea virus (PEDV) COE antigen was given to 0 day old and 2 day old piglets and they were challenged with infectious PEDV 5 days later. The vaccine triggered antigen-specific systemic and mucosal immune response, increased weight gain even after PEDV challenge and increased survival. Further this mucosally administered bacterial vaccine showed elevated transforming growth factor-β in the mesenteric lymph nodes suggesting induction of an anti-inflammatory during PEDV infection. Other research has shown that groups of one-day-old broiler chicks were vaccinated via the oculo-nasal route with different live infectious bronchitis virus (IBV) vaccines (123). After 21 days, the birds responded with increased CD4+, CD8+ and IgA-bearing B cells in the trachea and some evidence of ciliary protection against challenge (123). Mucosally administered vaccines are showing promise in the newborn animal.

Modulation of neonatal immunity

Probiotics

Use of probiotics may be a means to improve oral vaccine efficacy in neonates. For example, low doses of *Lactobacillus rhamnosus* GG (LGG) and *Bifidobacterium animalis* subsp. *lactis* Bb12 (Bb12), probiotics that mimic colonization by commensal species, were administered to gnotiobtioic piglets to enhance rotavirus vaccine efficacy (124). Gnotobiotic piglets colonized with probiotics and vaccinated thrice with attenuated rotavirus had reduced diarrhea severity and shedding scores compared with piglets not administered the probiotics. The probiotic-treated and challenged piglets had elevated intestinal IgA antibody responses, IFN-γ-producing $CD4^+$ and $CD8^+$ T cells, and serum Th1 cytokine responses post challenge compared with piglets vaccinated with only the attenuated RV vaccine (124). When the probiotics were combined with colostrum/milk supplementation, the vaccinated piglets showed enhanced anti-RV serum IgA titres and RV-specific IgA antibody-secreting cells in the piglet intestine (125). This study showed that appropriate commensals or probiotics may help to overcome the suppressive effects of low titers of maternal antibodies on neonatal vaccine-specific IgA antibody responses. Another study went one step further and showed that supplementation with probiotics promoted protection against infectious challenge. One-day old sow-reared piglets were allowed to suckle and then one group was orally supplemented with *Enterococcus faecium* NCIMB 10415 (*E. faecium*) daily for 1 week (126). On the 8th day, a subset of piglets were challenged with enterotoxigenic *Escherichia coli* K88 (ETEC). Supplementation with *E. faecium* partly alleviated the intestinal injury and diarrhea severity of neonatal piglets challenged by ETEC (126). Supplementation with commensal flora during the suckling period should be investigated further as a means to augment the neonatal immune response during this vulnerable period.

Immunostimulants

Vaccines that include immunostimulants may be a viable option to augment neonatal immunity. *M. bovis* BCG is a potent immunomodulatory agent. Foals are known to be deficient in production of IFNγ and IL-4 and therefore they may be at a disadvantage to older animals in mounting a vaccine or infection-specific T cell response (127,128). Four-day old foals were immunized subcutaneously with live M. bovis BCG and they responded with significant antigen-specific primary humoral immunity. Further, they showed significant induction of delayed type hypersensitivity (DTH) responses eight weeks after vaccination, as measured by the increase in double skin thickness after intradermal injection of purified protein derivative suggesting successful induction of a functional CMI (129). Others have used Mycobacterium Cell Wall Fraction (MCWF) as a biological immunomodulatory agent in suckling calves (130) They observed that calves supplemented with MCWF had significantly reduced ETEC disease severity and duration of diarrhea and reduced mortality compared to controls. Thus, whether M. bovis BCG or MCWF can be used in conjunction with vaccines and/or whether the M. bovis BCG can be used as a vaccine vector should be investigated to promote neonatal immunity in livestock species.

Other immunostimulants can also protect newborn animals against ETEC or other diseases. For instance, CpG ODN and host defense peptide, administered intranasally to 3-7-day old piglets was associated with increased protection against ETEC (131). One-day old goats were administered the lipophilic immunomodulator β-D-GlcNstearoyl-(1-4)-norMurNAc-L-Abu-D-isoGln (N-L18-norAbu-GMDP) via a combined intranasal and oral route 72 and 24 h pre-infection challenge. The kids were orally challenged at 6 days of age with 1×10^7 oocysts of *C. parvum*. Kids given the cumulative dose of 2000 µg immunostimulant per kid 72 and 24 h pre-infection challenge, lead to substantially increased clearance of coccidian parasites from various parts of the intestine (132). In other research, specific pathogen-free chickens were supplemented with a glycosylated antimicrobial pathogen peptide produced by *Bacillus subtilis* 168, at 14 days of age or they were administered no addition to their diet (133). Chickens were vaccinated with Newcastle disease (ND) via the intranasal and intraocular routes 2 weeks later. The vaccinated and supplemented birds had increased B lymphocyte proliferation and antibody titres and increased proportion of $CD8^+$ T lymphocyte subpopulations suggesting that the glycosylated antimicrobial pathogen peptide promoted both humoral and cellular immune responses and may be an effective vaccine adjuvant (133). Thus, giving newborn animals immunostimulants may bolster their protection against early infectious challenges and should be investigated for livestock or companion animals where vaccines are not currently available.

Micronutrients

Several micronutrients may be lacking in colostrum and milk and supplementation to the suckling offspring may improve their health and ability to respond to vaccines. For instance, preweaning calves from an area in Argentina with known Cu and Zn deficiencies were supplemented with Cu and Zn. Calves were then vaccinated with inactivated bovine herpes virus-1 vaccine. Supplemented animals had higher overall weight and responded with significantly elevated anti-BoHV1 immunoglobulins (134). It remains to be proven whether their addition to the diet may improve vaccine efficacy but it is a strategy that should be investigated further for high-value animals.

Conclusions

Neonatal vaccination is a critical element in the management of livestock and companion animals. However, many neonatal vaccines are ineffective due to a failure to address the unique features of the neonatal immune system. As our understanding of the differences between the neonatal and adult immune system continue to develop, improved vaccine formulations will become available that offer higher effectiveness for this age group combined with improved stability and reduced cost. Similarly, vaccine formulations will become available for special populations including older animals or even pregnant animals. Such precision vaccines will

Table 2. Strategies to Promote Neonatal Immunity

Strategy	Approach	Mechanism
Vaccines	Vectored-vaccines	Effective induction of Th-1 responses, help to overcome interference with maternal immunity
	Nucleic-acid vaccines	Effective induction of humoral and cell-mediated responses in the neonate; or interference with maternal antibodies
	Particle-based vaccines	Hide antigen from maternal antibodies and promote uptake by antigen-presenting cells
	Adjuvants	Drive the immune response in neonates and ensure early onset and long lasting immunity
Immune modulation	Probiotics	Can be used to stimulate the neonatal immune system and in particular activate neonatal antigen presenting cells
	Micronutrients and Immune modulators	Bolster their protection and stimulate the neonatal immune system
Alternative routes of immunization	Target mucosal sites	Trigger immunity in mucosal compartments as well as systemically

become an economically important disease management tool for veterinary applications, and in combination with maternal vaccines, can be used to protect newborn animals against infectious diseases. Ideally, a combination of affordable passive immunization of the dam and neonatal immunization would be jointly used to protect the vulnerable young population against infectious diseases to improve animal health, welfare and return on investment for producers.

References

1. Banks, K. L. (1981) Changes in the immune response related to age. *Vet Clin North Am Small Anim Pract* **11**, 683-688
2. Djonne, B. (2007) Infections and perinatal diseases - a comparative overview. *Acta Vet Scand* **49**
3. Martin, B. R., and Larson, K. A. (1973) Immune response of equine fetus to coliphage T2. *Am J Vet Res* **34**, 1363-1364
4. Sarma, G., Greer, W., Gildersleeve, R. P., Murray, D. L., and Miles, A. M. (1995) Field safety and efficacy of in ovo administration of HVT + SB-1 bivalent Marek's disease vaccine in commercial broilers. *Avian Dis* **39**, 211-217
5. Reddy, S. K., Sharma, J. M., Ahmad, J., Reddy, D. N., McMillen, J. K., Cook, S. M., Wild, M. A., and Schwartz, R. D. (1996) Protective efficacy of a

recombinant herpesvirus of turkeys as an in ovo vaccine against Newcastle and Marek's diseases in specific-pathogen-free chickens. *Vaccine* **14**, 469-477

6. Sharma, J. M. (1986) Embryo vaccination of specific-pathogen-free chickens with infectious bursal disease virus: tissue distribution of the vaccine virus and protection of hatched chickens against disease. *Avian Dis* **30**, 776-780
7. Ding, X., Lillehoj, H. S., Quiroz, M. A., Bevensee, E., and Lillehoj, E. P. (2004) Protective immunity against Eimeria acervulina following in ovo immunization with a recombinant subunit vaccine and cytokine genes. *Infect Immun* **72**, 6939-6944
8. Gerdts, V., Babiuk, L. A., Littel-van den Hurk, S. V., and Griebel, P. J. (2000) Fetal immunization by a DNA vaccine delivered into the oral cavity. *Nat Med* **6**, 929-932
9. Gerdts, V., Snider, M., Brownlie, R., Babiuk, L. A., and Griebel, P. J. (2002) Oral DNA vaccination in utero induces mucosal immunity and immune memory in the neonate. *J Immunol* **168**, 1877-1885
10. Harbeson, D., Ben-Othman, R., Amenyogbe, N., and Kollmann, T. R. (2018) Outgrowing the Immaturity Myth: The Cost of Defending From Neonatal Infectious Disease. *Frontiers in immunology* **9**
11. Ott, T. L. (2020) Immunological detection of pregnancy: Evidence for systemic immune modulation during early pregnancy in ruminants. *Theriogenology* **150**, 498-503
12. Yu, J. C., Khodadadi, H., Malik, A., Davidson, B., Salles, E. D. L., Bhatia, J., Hale, V. L., and Baban, B. (2018) Innate Immunity of Neonates and Infacts. *Frontiers in immunology* **9**
13. Kollmann, T. R., Marchant, A., and Way, S. S. (2020) Vaccination strategies to enhance immunity in neonates. *Science* **368**, 612-615
14. Kumar, S. K. M., and Bhat, B. V. (2016) Distinct mechanisms of the newborn innate immunity. *Immunol Lett* **173**, 42-54
15. Simmerman, E., Qin, X., Marshall, B., Perry, L., Cai, L., Wang, T., Yu, J., Akbari, O., and Baban, B. (2016) Innate lymphoid cells: a paradigm for low SSI in cleft lip repair. *J Surg Res* **205**, 312-317
16. Oherle, K., Acker, E., Bonfield, M., Wang, T., Gray, J., Lang, I., Bridges, J., Lewkowich, I., Xu, Y., Ahlfeld, S., Zacharias, W., Alenghat, T., and Deshmukh, H. (2020) Insulin-like Growth Factor 1 Supports a Pulmonary Niche that

Promotes Type 3 Innate Lymphoid Cell Development in Newborn Lungs. *Immunity* **52**, 275-+

17. van Splunter, M., van Osch, T. L. J., Brugman, S., Savelkoul, H. F. J., Joosten, L. A. B., Netea, M. G., and van Neerven, R. J. J. (2018) Induction of Trained Innate Immunity in Human Monocytes by Bovine Milk and Milk-Derived Immunoglobulin G. *Nutrients* **10**, 1378
18. Guerra-Maupome, M., Vang, D. X., and McGill, J. L. (2019) Aerosol vaccination with Bacille Calmette-Guerin induces a trained innate immune phenotype in calves. *PloS one* **14**, e0212751
19. Angulo, M., Reyes-Becerril, M., Cepeda-Palacios, R., and Angulo, C. (2020) Oral administration of Debaryomyces hansenii CBS8339-β-glucan induces trained immunity in newborn goats. *Dev Comp Immunol* **105**, 103597
20. Byrne, K. A., Loving, C. L., and McGill, J. L. (2020) Innate Immunomodulation in Food Animals: Evidence for Trained Immunity? *Frontiers in immunology* **11**
21. Mastelic-Gavillet, B., Vono, M., Gonzalez-Dias, P., Ferreira, F. M., Cardozo, L., Lambert, P. H., Nakaya, H. I., and Siegrist, C. A. (2019) Neonatal T Follicular Helper Cells Are Lodged in a Pre-T Follicular Helper Stage Favoring Innate Over Adaptive Germinal Center Responses. *Frontiers in immunology* **10**
22. Pihlgren, M., Tougne, C., Bozzotti, P., Fulurija, A., Duchosal, M. A., Lambert, P. H., and Siegrist, C. A. (2003) Unresponsiveness to lymphoid-mediated signals at the neonatal follicular dendritic cell precursor level contributes to delayed germinal center induction and limitations of neonatal antibody responses to T-dependent antigens. *J Immunol* **170**, 2824-2832
23. Mastelic, B., Kamath, A. T., Fontannaz, P., Tougne, C., Rochat, A. F., Belnoue, E., Combescure, C., Auderset, F., Lambert, P. H., Tacchini-Cottier, F., and Siegrist, C. A. (2012) Environmental and T Cell-Intrinsic Factors Limit the Expansion of Neonatal Follicular T Helper Cells but May Be Circumvented by Specific Adjuvants. *J Immunol* **189**, 5764-5772
24. Debock, I., Jaworski, K., Chadlaoui, H., Delbauve, S., Passon, N., Twyffels, L., Leo, O., and Flamand, V. (2013) Neonatal Follicular Th Cell Responses Are Impaired and Modulated by IL-4. *J Immunol* **191**, 1231-1239
25. Butler, J. E., Santiago-Mateo, K., Wertz, N., Sun, X. Z., Sinkora, M., and Francis, D. L. (2016) Antibody repertoire development in fetal and neonatal piglets. XXIV. Hypothesis: The ileal Peyer patches (IPP) are the major source

of primary, undiversified IgA antibodies in newborn piglets. *Dev Comp Immunol* **65**, 340-351

26. Chattha, K. S., Firth, M. A., Hodgins, D. C., and Shewen, P. E. (2009) Age related variation in expression of CD21 and CD32 on bovine lymphocytes: a cross-sectional study. *Veterinary immunology and immunopathology* **130**, 70-78
27. Mccauley, I., and Hartmann, P. E. (1984) Changes in Piglet Leukocytes, Lymphocytes-B and Plasma-Cortisol from Birth to 3 Weeks after Weaning. *Res Vet Sci* **37**, 234-241
28. Baintner, K. (2007) Transmission of antibodies from mother to young: Evolutionary strategies in a proteolytic environment. *Veterinary immunology and immunopathology* **117**, 153-161
29. Westrom, B. R., Svendsen, J., Ohlsson, B. G., Tagesson, C., and Karlsson, B. W. (1984) Intestinal transmission of macromolecules (BSA and FITC-labelled dextrans) in the neonatal pig. Influence of age of piglet and molecular weight of markers. *Biol Neonate* **46**, 20-26
30. Tyler, H., and Ramsey, H. (1993) Effect of insulin-induced hypoglycemia on cessation of macromolecular transport in the neonatal calf. *J Dairy Sci* **76**, 2736-2741
31. Hedegaard, C. J., and Heegaard, P. M. H. (2016) Passive immunisation, an old idea revisited: Basic principles and application to modern animal production systems. *Veterinary immunology and immunopathology* **174**, 50-63
32. Windeyer, M. C., Leslie, K. E., Godden, S. M., Hodgins, D. C., Lissemore, K. D., and LeBlanc, S. J. (2014) Factors associated with morbidity, mortality, and growth of dairy heifer calves up to 3 months of age. *Prev Vet Med* **113**, 231-240
33. Godden, S. M., Lombard, J. E., and Woolums, A. R. (2019) Colostrum Management for Dairy Calves. *Vet Clin North Am Food Anim Pract* **35**, 535-556
34. Fischer, A. J., Song, Y., He, Z., Haines, D. M., Guan, L. L., and Steele, M. A. (2018) Effect of delaying colostrum feeding on passive transfer and intestinal bacterial colonization in neonatal male Holstein calves. *J Dairy Sci* **101**, 3099-3109
35. Dauvergne, M., Brun, A., and Soulebot, J. P. (1983) Passive protection of newborn calves against rotavirus by vaccination of their dams. *Dev Biol Stand* **53**, 245-255

36. Jayappa, H., Davis, R., Dierks, L., Sweeney, D., and Wasmoen, T. (2008) Demonstration of passive protection in neonatal calves against colibacillosis following immunization of pregnant heifers at 3 months of gestation. *Vet Ther* **9**, 283-289
37. Mukkur, T. K., Walker, K. H., and McDowell, G. H. (1998) Passive immunisation of neonatal lambs via colostrum and milk of ewes previously immunised with live attenuated Salmonella typhimurium protects neonatal lambs from experimental salmonellosis. *Comp Immunol Microbiol Infect Dis* **21**, 327-336
38. Altmann, K., and Mukkur, T. K. (1983) Passive immunisation of neonatal lambs against infection with enteropathogenic Escherichia coli via colostrum of ewes immunised with crude and purified K99 pili. *Res Vet Sci* **35**, 234-239
39. Poffenbarger, E. M., Olson, P. N., Chandler, M. L., Seim, H. B., and Varman, M. (1991) Use of adult dog serum as a substitute for colostrum in the neonatal dog. *American journal of veterinary research* **52**, 1221-1224
40. Mila, H., Grellet, A., Desario, C., Feugier, A., Decaro, N., Buonavoglia, C., and Chastant-Maillard, S. (2014) Protection against canine parvovirus type 2 infection in puppies by colostrum-derived antibodies. *J Nutr Sci* **3**, e54
41. Lecce, J. G. (1973) Effect of Dietary Regimen on Cessation of Uptake of Macromolecules by Piglet Intestinal Epithelium (Closure) and Transport to Blood. *J Nutr* **103**, 751-756
42. Kowalczyk, K., Daiss, J., Halpern, J., and Roth, T. F. (1985) Quantitation of maternal-fetal IgG transport in the chicken. *Immunology* **54**, 755-762
43. Hamal, K. R., Burgess, S. C., Pevzner, I. Y., and Erf, G. F. (2006) Maternal antibody transfer from dams to their egg yolks, egg whites, and chicks in meat lines of chickens. *Poult Sci* **85**, 1364-1372
44. Gupta, A., Popowich, S., Ojkic, D., Kurukulasuriya, S., Chow-Lockerbie, B., Gunawardana, T., Goonewardene, K., Karunarathna, R., Ayalew, L. E., Ahmed, K. A., Tikoo, S. K., Willson, P., and Gomis, S. (2018) Inactivated and live bivalent fowl adenovirus (FAdV8b + FAdV11) breeder vaccines provide broad-spectrum protection in chicks against inclusion body hepatitis (IBH). *Vaccine* **36**, 744-750
45. Swain, P., and Nayak, S. K. (2009) Role of maternally derived immunity in fish. *Fish Shellfish Immunol* **27**, 89-99

46. Sin, Y. M., Ling, K. H., and Lam, T. J. (1994) Passive transfer of protective immunity against ichthyophthiriasis from vaccinated mother to fry in tilapias, Oreochromis aureus. *Aquaculture* **120**, 229-237
47. Oshima, S., Hata, J., Segawa, C., and Yamashita, S. (1996) Mother to fry, successful transfer of immunity against infectious haematopoietic necrosis virus infection in rainbow trout. *J Gen Virol* **77 (Pt 10)**, 2441-2445
48. Langel, S. N., Paim, F. C., Alhamo, M. A., Buckley, A., Van Geelen, A., Lager, K. M., Vlasova, A. N., and Saif, L. J. (2019) Stage of Gestation at Porcine Epidemic Diarrhea Virus Infection of Pregnant Swine Impacts Maternal Immunity and Lactogenic Immune Protection of Neonatal Suckling Piglets. *Frontiers in immunology* **10**, 727
49. Langel, S. N., Paim, F. C., Alhamo, M. A., Lager, K. M., Vlasova, A. N., and Saif, L. J. (2019) Oral vitamin A supplementation of porcine epidemic diarrhea virus infected gilts enhances IgA and lactogenic immune protection of nursing piglets. *Veterinary research* **50**, 101
50. Wen, Z., Xu, Z., Zhou, Q., Li, W., Wu, Y., Du, Y., Chen, L., Xue, C., and Cao, Y. (2019) A heterologous 'prime-boost' anti-PEDV immunization for pregnant sows protects neonatal piglets through lactogenic immunity against PEDV. *Letters in applied microbiology* **69**, 258-263
51. Sareyyüpoğlu, B., Gülyaz, V., Çokçalışkan, C., Ünal, Y., Çökülgen, T., Uzunlu, E., Gürcan, S., and İlk, O. (2019) Effect of FMD vaccination schedule of dams on the level and duration of maternally derived antibodies. *Veterinary immunology and immunopathology* **217**, 109881
52. Zhang, J., Chen, J., Liu, Y., Da, S., Shi, H., Zhang, X., Liu, J., Cao, L., Zhu, X., Wang, X., Ji, Z., and Feng, L. (2020) Pathogenicity of porcine deltacoronavirus (PDCoV) strain NH and immunization of pregnant sows with an inactivated PDCoV vaccine protects 5-day-old neonatal piglets from virulent challenge. *Transbound Emerg Dis* **67**, 572-583
53. Liebler-Tenorio, E. M., Riedel-Caspari, G., and Pohlenz, J. F. (2002) Uptake of colostral leukocytes in the intestinal tract of newborn calves. *Veterinary immunology and immunopathology* **85**, 33-40
54. Tuboly, S., Bernáth, S., Glávits, R., and Medveczky, I. (1988) Intestinal absorption of colostral lymphoid cells in newborn piglets. *Veterinary immunology and immunopathology* **20**, 75-85

55. Bandrick, M., Ariza-Nieto, C., Baidoo, S. K., and Molitor, T. W. (2014) Colostral antibody-mediated and cell-mediated immunity contributes to innate and antigen-specific immunity in piglets. *Dev & Comp Immunol* **43**, 114-120
56. Park, Y. H., Fox, L. K., Hamilton, M. J., and Davis, W. C. (1992) Bovine mononuclear leukocyte subpopulations in peripheral blood and mammary gland secretions during lactation. *J Dairy Sci* **75**, 998-1006
57. Reber, A. J., Donovan, D. C., Gabbard, J., Galland, K., Aceves-Avila, M., Holbert, K. A., Marshall, L., and Hurley, D. J. (2008) Transfer of maternal colostral leukocytes promotes development of the neonatal immune system Part II. Effects on neonatal lymphocytes. *Veterinary immunology and immunopathology* **123**, 305-313
58. Reber, A. J., Hippen, A. R., and Hurley, D. J. (2005) Effects of the ingestion of whole colostrum or cell-free colostrum on the capacity of leukocytes in newborn calves to stimulate or respond in one-way mixed leukocyte cultures. *Am J Vet Res* **66**, 1854-1860
59. Baban, B., Malik, A., Bhatia, J., and Yu, J. C. (2018) Presence and Profile of Innate Lymphoid Cells in Human Breast Milk. *JAMA Pediatr* **172**, 594-596
60. Stelwagen, K., Carpenter, E., Haigh, B., Hodgkinson, A., and Wheeler, T. T. (2009) Immune components of bovine colostrum and milk. *Journal of animal science* **87**, 3-9
61. Joslin, R. S., Erickson, P. S., Santoro, H. M., Whitehouse, N. L., Schwab, C. G., and Rejman, J. J. (2002) Lactoferrin supplementation to dairy calves. *J Dairy Sci* **85**, 1237-1242
62. Wheeler, T. T., Smolenski, G. A., Harris, D. P., Gupta, S. K., Haigh, B. J., Broadhurst, M. K., Molenaar, A. J., and Stelwagen, K. (2011) Host-defence-related proteins in cows' milk. *Animal* **6**, 415-422
63. Haisan, J., Oba, M., Ambrose, D. J., and Steele, M. A. (2018) Short communication: The effects of offering a high or low plane of milk preweaning on insulin-like growth factor and insulin-like growth factor binding proteins in dairy heifer calves. *J Dairy Sci* **101**, 11441-11446
64. Woliński, J., Słupecka, M., and Romanowicz, K. (2014) Leptin and ghrelin levels in colostrum, milk and blood plasma of sows and pig neonates during the first week of lactation. *Animal Science Journal* **85**, 143-149

65. Nissen, A., Andersen, P. H., Bendixen, E., Ingvartsen, K. L., and Røntved, C. M. (2017) Colostrum and milk protein rankings and ratios of importance to neonatal calf health using a proteomics approach. *J Dairy Sci* **100**, 2711-2728
66. Hedegaard, C. J., and Heegaard, P. M. (2016) Passive immunisation, an old idea revisited: Basic principles and application to modern animal production systems. *Veterinary immunology and immunopathology* **174**, 50-63
67. Kovacs-Nolan, J., and Mine, Y. (2012) Egg yolk antibodies for passive immunity. *Annu Rev Food Sci Technol* **3**, 163-182
68. Lora, I., Gottardo, F., Bonfanti, L., Stefani, A. L., Soranzo, E., Dall'Ava, B., Capello, K., Martini, M., and Barberio, A. (2019) Transfer of passive immunity in dairy calves: the effectiveness of providing a supplementary colostrum meal in addition to nursing from the dam. *Animal* **13**, 2621-2629
69. Marquardt, R. R., Jin, L. Z., Kim, J. W., Fang, L., Frohlich, A. A., and Baidoo, S. K. (1999) Passive protective effect of egg-yolk antibodies against enterotoxigenic Escherichia coli K88+ infection in neonatal and early-weaned piglets. *FEMS Immunol Med Microbiol* **23**, 283-288
70. Vega, C., Bok, M., Chacana, P., Saif, L., Fernandez, F., and Parreno, V. (2011) Egg yolk IgY: protection against rotavirus induced diarrhea and modulatory effect on the systemic and mucosal antibody responses in newborn calves. *Veterinary immunology and immunopathology* **142**, 156-169
71. Ikemori, Y., Ohta, M., Umeda, K., Icatlo, F. C., Jr., Kuroki, M., Yokoyama, H., and Kodama, Y. (1997) Passive protection of neonatal calves against bovine coronavirus-induced diarrhea by administration of egg yolk or colostrum antibody powder. *Vet Microbiol* **58**, 105-111
72. Mila, H., Grellet, A., Mariani, C., Feugier, A., Guard, B., Suchodolski, J., Steiner, J., and Chastant-Maillard, S. (2017) Natural and artificial hyperimmune solutions: Impact on health in puppies. *Reprod Domest Anim* **52 Suppl 2**, 163-169
73. Gao, X., Zhang, X., Lin, L., Yao, D., Sun, J., Du, X., Li, X., and Zhang, Y. (2016) Passive Immune-Protection of Litopenaeus vannamei against Vibrio harveyi and Vibrio parahaemolyticus Infections with Anti-Vibrio Egg Yolk (IgY)-Encapsulated Feed. *Int J Mol Sci* **17**
74. Le Bourgot, C., Le Normand, L., Formal, M., Respondek, F., Blat, S., Apper, E., Ferret-Bernard, S., and Le Huerou-Luron, I. (2017) Maternal short-chain fructo-oligosaccharide supplementation increases intestinal cytokine secretion, goblet

cell number, butyrate concentration and Lawsonia intracellularis humoral vaccine response in weaned pigs. *Br J Nutr* **117**, 83-92

75. Bedenice, D. (2019) Overview of Management of the Neonate in Large Animals. in *Merck Veterinary Manual*, MERCK & CO., Inc., Kenilworth, NJ, USA
76. Haverson, K., Corfield, G., Jones, P. H., Kenny, M., Fowler, J., Bailey, M., Stokes, C. R., and Miller, B. G. (2009) Effect of oral antigen and antibody exposure at birth on subsequent immune status. A study in neonatal pigs. *International archives of allergy and immunology* **150**, 192-204
77. Vono, M., Eberhardt, C. S., Auderset, F., Mastelic-Gavillet, B., Lemeille, S., Christensen, D., Andersen, P., Lambert, P. H., and Siegrist, C. A. (2019) Maternal Antibodies Inhibit Neonatal and Infant Responses to Vaccination by Shaping the Early-Life B Cell Repertoire within Germinal Centers. *Cell Rep* **28**, 1773-+
78. Kim, D., Huey, D., Oglesbee, M., and Niewiesk, S. (2011) Insights into the regulatory mechanism controlling the inhibition of vaccine-induced seroconversion by maternal antibodies. *Blood* **117**, 6143-6151
79. Chamorro, M. F., Woolums, A., and Walz, P. H. (2016) Vaccination of calves against common respiratory viruses in the face of maternally derived antibodies (IFOMA). *Anim Health Res Rev* **17**, 79-84
80. Windeyer, M. C., and Gamsjager, L. (2019) Vaccinating Calves in the Face of Maternal Antibodies Challenges and Opportunities. *Vet Clin N Am-Food A* **35**, 557-+
81. Gans, H. A., Yasukawa, L. L., Alderson, A., Rinki, M., DeHovitz, R., Beeler, J., Audet, S., Maldonado, Y., and Arvin, A. M. (2004) Humoral and cell-mediated immune responses to an early 2-dose measles vaccination regimen in the United States. *J Infect Dis* **190**, 83-90
82. Gans, H. A., Maldonado, Y., Yasukawa, L. L., Beeler, J., Audet, S., Rinki, M. M., DeHovitz, R., and Arvin, A. M. (1999) IL-12, IFN-gamma, and T cell proliferation to measles in immunized infants. *J Immunol* **162**, 5569-5575
83. Pitcovski, J., Pitcovski, E., Goldenberg, D., and Shahar, E. (2017) Pair-epitopes vaccination: enabling offspring vaccination in the presence of maternal antibodies. *Avian Pathol* **46**, 581-584

84. de Bree, L. C. J., Koeken, V. A. C. M., Joosten, L. A. B., Aaby, P., Benn, C. S., van Crevel, R., and Netea, M. G. (2018) Non-specific effects of vaccines: Current evidence and potential implications. *Semin Immunol* **39**, 35-43
85. Goodridge, H. S., Ahmed, S. S., Curtis, N., Kollmann, T. R., Levy, O., Netea, M. G., Pollard, A. J., van Crevel, R., and Wilson, C. B. (2016) Harnessing the beneficial heterologous effects of vaccination. *Nat Rev Immunol* **16**, 392-400
86. Ellis, J. A., Gow, S. P., Mahan, S., and Leyh, R. (2013) Duration of immunity to experimental infection with bovine respiratory syncytial virus following intranasal vaccination of young passively immune calves. *J Am Vet Med Assoc* **243**, 1602-1608
87. Ellis, J. A. (2017) How efficacious are vaccines against bovine respiratory syncytial virus in cattle? *Vet Microbiol* **206**, 59-68
88. Genzow, M., Goodell, C., Kaiser, T. J., Johnson, W., and Eichmeyer, M. (2018) Live attenuated influenza virus vaccine reduces virus shedding of newborn piglets in the presence of maternal antibody. *Influenza and other respiratory viruses* **12**, 353-359
89. Hill, K. L., Hunsaker, B. D., Townsend, H. G., van Drunen Littel-van den Hurk, S., and Griebel, P. J. (2012) Mucosal immune response in newborn Holstein calves that had maternally derived antibodies and were vaccinated with an intranasal multivalent modified-live virus vaccine. *J Am Vet Med Assoc* **240**, 1231-1240
90. Jeong, J., Kim, S., Park, K. H., Kang, I., Park, S.-J., Yang, S., Oh, T., and Chae, C. (2018) Vaccination with a porcine reproductive and respiratory syndrome virus vaccine at 1-day-old improved growth performance of piglets under field conditions. *Vet Microbiol* **214**, 113-124
91. Tan, J., Cooke, J., Clarke, N., and Tannock, G. A. (2007) Molecular evaluation of responses to vaccination and challenge by Marek's disease viruses. *Avian Pathol* **36**, 351-359
92. Pardo, M. C., Tanner, P., Bauman, J., Silver, K., and Fischer, L. (2007) Immunization of puppies in the presence of maternally derived antibodies against canine distemper virus. *J Comp Pathol* **137**, S72-S75
93. Fougerolle, S., Legrand, L., Garrett, D., Birand, I., Foursin, M., D'Ablon, X., Bayssat, P., Newton, R. J., Pronost, S., and Paillot, R. (2016) Influential factors inducing suboptimal humoral response to vector-based influenza immunisation in Thoroughbred foals. *Vaccine* **34**, 3787-3795

94. Welter, J., Taylor, J., Tartaglia, J., Paoletti, E., and Stephensen, C. B. (2000) Vaccination against canine distemper virus infection in infant ferrets with and without maternal antibody protection, using recombinant attenuated poxvirus vaccines. *J Virol* **74**, 6358-6367

95. Pedrera, M., Macchi, F., McLean, R. K., Franceschi, V., Thakur, N., Russo, L., Medfai, L., Todd, S., Tchilian, E. Z., Audonnet, J. C., Chappell, K., Isaacs, A., Watterson, D., Young, P. R., Marsh, G. A., Bailey, D., Graham, S. P., and Donofrio, G. (2020) Bovine Herpesvirus-4-Vectored Delivery of Nipah Virus Glycoproteins Enhances T Cell Immunogenicity in Pigs. *Vaccines-Basel* **8**

96. Kyriakis, C. S., De Vleeschauwer, A., Barbe, F., Bublot, M., and Van Reeth, K. (2009) Safety, immunogenicity and efficacy of poxvirus-based vector vaccines expressing the haemagglutinin gene of a highly pathogenic H5N1 avian influenza virus in pigs. *Vaccine* **27**, 2258-2264

97. Monteil, M., Le Pottier, M. F., Ambriović Ristov, A., Cariolet, R., L'Hospitalier, R., Klonjkowski, B., and Eloit, M. (2000) Single inoculation of replication-defective adenovirus-vectored vaccines at birth in piglets with maternal antibodies induces high level of antibodies and protection against pseudorabies. *Vaccine* **18**, 1738-1742

98. Minke, J. M., Toulemonde, C. E., Dinic, S., Cozette, V., Cullinane, A., and Audonnet, J. C. (2007) Effective priming of foals born to immune dams against influenza by a canarypox-vectored recombinant influenza H3N8 vaccine. *J Comp Pathol* **137**, S76-S80

99. Wang, J., Huang, L., Mou, C., Zhang, E., Wang, Y., Cao, Y., and Yang, Q. (2019) Mucosal immune responses induced by oral administration recombinant Bacillus subtilis expressing the COE antigen of PEDV in newborn piglets. *Biosci Rep* **39**, BSR20182028

100. Ding, X., Lillehoj, H. S., Quiroz, M. A., Bevensee, E., and Lillehoj, E. P. (2004) Protective immunity against Eimeria acervulina following in ovo immunization with a recombinant subunit vaccine and cytokine genes. *Infect Immun* **72**, 6939-6944

101. Dorosti, H., Eslami, M., Negahdaripour, M., Ghoshoon, M. B., Gholami, A., Heidari, R., Dehshahri, A., Erfani, N., Nezafat, N., and Ghasemi, Y. (2019) Vaccinomics approach for developing multi-epitope peptide pneumococcal vaccine. *Journal of biomolecular structure & dynamics* **37**, 3524-3535

102. Faburay, B., Lebedev, M., McVey, D. S., Wilson, W., Morozov, I., Young, A., and Richt, J. A. (2014) A Glycoprotein Subunit Vaccine Elicits a Strong Rift Valley Fever Virus Neutralizing Antibody Response in Sheep. *Vector-Borne Zoonot* **14**, 746-756
103. Plummer, E. M., and Manchester, M. (2011) Viral nanoparticles and virus-like particles: platforms for contemporary vaccine design. *Wiley interdisciplinary reviews. Nanomedicine and nanobiotechnology* **3**, 174-196
104. Souza, M., Costantini, V., Azevedo, M. S., and Saif, L. J. (2007) A human norovirus-like particle vaccine adjuvanted with ISCOM or mLT induces cytokine and antibody responses and protection to the homologous GII.4 human norovirus in a gnotobiotic pig disease model. *Vaccine* **25**, 8448-8459
105. Han, M. G., Cheetham, S., Azevedo, M., Thomas, C., and Saif, L. J. (2006) Immune responses to bovine norovirus-like particles with various adjuvants and analysis of protection in gnotobiotic calves. *Vaccine* **24**, 317-326
106. Dhakal, S., and Renukaradhya, G. J. (2019) Nanoparticle-based vaccine development and evaluation against viral infections in pigs. *Veterinary research* **50**, 90
107. McGill, J. L., Kelly, S. M., Kumar, P., Speckhart, S., Haughney, S. L., Henningson, J., Narasimhan, B., and Sacco, R. E. (2018) Efficacy of mucosal polyanhydride nanovaccine against respiratory syncytial virus infection in the neonatal calf. *Sci Rep* **8**, 3021
108. Tran, T.-L., Castagne, N., Bhella, D., Varela, P. F., Bernard, J., Chilmonczyk, S., Berkenkamp, S., Benhamo, V., Grznarova, K., Grosclaude, J., Nespoulos, C., Rey, F. A., and Eleouet, J.-F. (2007) The nine C-terminal amino acids of the respiratory syncytial virus protein P are necessary and sufficient for binding to ribonucleoprotein complexes in which six ribonucleotides are contacted per N protein protomer. *J Gen Virol* **88**, 196-206
109. Riffault, S., Meyer, G., Deplanche, M., Dubuquoy, C., Durand, G., Soulestin, M., Castagné, N., Bernard, J., Bernardet, P., Dubosclard, V., Bernex, F., Petit-Camurdan, A., Deville, S., Schwartz-Cornil, I., and Eléouët, J.-F. (2010) A new subunit vaccine based on nucleoprotein nanoparticles confers partial clinical and virological protection in calves against bovine respiratory syncytial virus. *Vaccine* **28**, 3722-3734
110. Schnee, M., Vogel, A. B., Voss, D., Petsch, B., Baumhof, P., Kramps, T., and Stitz, L. (2016) An mRNA Vaccine Encoding Rabies Virus Glycoprotein Induces

Protection against Lethal Infection in Mice and Correlates of Protection in Adult and Newborn Pigs. *PLoS neglected tropical diseases* **10**, e0004746

111. Hamers, C., Juillard, V., and Fischer, L. (2007) DNA Vaccination against Pseudorabies Virus and Bovine Respiratory Syncytial Virus Infections of Young Animals in the Face of Maternally Derived Immunity. *J Comp Biol* **137**, S35-S41
112. Young, A. J. (2019) Adjuvants What a Difference 15 Years Makes! *Vet Clin N Am-Food A* **35**, 391-+
113. Garg, R., Babiuk, L., Littel-van den Hurk, S. V., and Gerdts, V. (2017) A novel combination adjuvant platform for human and animal vaccines. *Vaccine* **35**, 4486-4489
114. Gerdts, V. (2015) Adjuvants for veterinary vaccines - types and modes of action. *Berl Munch Tierarztl* **128**, 456-463
115. Desclozeaux, M., Robbins, A., Jelocnik, M., Khan, S. A., Hanger, J., Gerdts, V., Potter, A., Polkinghorne, A., and Timms, P. (2017) Immunization of a wild koala population with a recombinant Chlamydia pecorum Major Outer Membrane Protein (MOMP) or Polymorphic Membrane Protein (PMP) based vaccine: New insights into immune response, protection and clearance. *Plos One* **12**
116. Allan, B., Wheler, C., Koster, W., Sarfraz, M., Potter, A., Gerdts, V., and Dar, A. (2018) In Ovo Administration of Innate Immune Stimulants and Protection from Early Chick Mortalities due to Yolk Sac Infection. *Avian Dis* **62**, 316-321
117. Auray, G., Facci, M. R., van Kessel, J., Buchanan, R., Babiuk, L. A., and Gerdts, V. (2013) Porcine Neonatal Blood Dendritic Cells, but Not Monocytes, Are More Responsive to TLRs Stimulation than Their Adult Counterparts. *Plos One* **8**
118. Gerdts, V., Littel-van den Hurk, S. V., and Potter, A. (2016) Protection of neonates and infants by maternal immunization. *Expert Rev Vaccines* **15**, 1347-1349
119. Polewicz, M., Gracia, A., Buchanan, R., Strom, S., Halperin, S. A., Potter, A. A., Babiuk, L. A., and Gerdts, V. (2011) Influence of maternal antibodies on active pertussis toxoid immunization of neonatal mice and piglets. *Vaccine* **29**, 7718-7726
120. Schulze, K., Ebensen, T., Babiuk, L. A., Gerdts, V., and Guzman, C. A. (2017) Intranasal vaccination with an adjuvanted polyphosphazenes nanoparticle-

based vaccine formulation stimulates protective immune responses in mice. *Nanomed-Nanotechnol* **13**, 2169-2178

121. Mutwiri, G. (2012) TLR9 agonists: Immune mechanisms and therapeutic potential in domestic animals. *Veterinary immunology and immunopathology* **148**, 85-89
122. Mutwiri, G., Bateman, C., Baca-Estrada, M. E., Snider, M., and Griebel, P. (2000) Induction of immune responses in newborn lambs following enteric immunization with a human adenovirus vaccine vector. *Vaccine* **19**, 1284-1293
123. Awad, F., Hutton, S., Forrester, A., Baylis, M., and Ganapathy, K. (2016) Heterologous live infectious bronchitis virus vaccination in day-old commercial broiler chicks: clinical signs, ciliary health, immune responses and protection against variant infectious bronchitis viruses. *Avian Pathol* **45**, 169-177
124. Chattha, K. S., Vlasova, A. N., Kandasamy, S., Rajashekara, G., and Saif, L. J. (2013) Divergent immunomodulating effects of probiotics on T cell responses to oral attenuated human rotavirus vaccine and virulent human rotavirus infection in a neonatal gnotobiotic piglet disease model. *J Immunol* **191**, 2446-2456
125. Chattha, K. S., Vlasova, A. N., Kandasamy, S., Esseili, M. A., Siegismund, C., Rajashekara, G., and Saif, L. J. (2013) Probiotics and colostrum/milk differentially affect neonatal humoral immune responses to oral rotavirus vaccine. *Vaccine* **31**, 1916-1923
126. Peng, X., Wang, R., Hu, L., Zhou, Q., Liu, Y., Yang, M., Fang, Z., Lin, Y., Xu, S., Feng, B., Li, J., Jiang, X., Zhuo, Y., Li, H., Wu, D., and Che, L. (2019) Enterococcus faecium NCIMB 10415 administration improves the intestinal health and immunity in neonatal piglets infected by enterotoxigenic Escherichia coli K88. *Journal of animal science and biotechnology* **10**, 72
127. Breathnach, C. C., Sturgill-Wright, T., Stiltner, J. L., Adams, A. A., Lunn, D. P., and Horohov, D. W. (2006) Foals are interferon gamma-deficient at birth. *Veterinary immunology and immunopathology* **112**, 199-209
128. Ryan, C., Giguere, S., Hagen, J., Hartnett, C., and Kalyuzhny, A. E. (2010) Effect of age and mitogen on the frequency of interleukin-4 and interferon gamma secreting cells in foals and adult horses as assessed by an equine-specific ELISPOT assay. *Veterinary immunology and immunopathology* **133**, 66-71

129. Sturgill, T. L., Giguere, S., Berghaus, L. J., Hurley, D. J., and Hondalus, M. K. (2014) Comparison of antibody and cell-mediated immune responses of foals and adult horses after vaccination with live Mycobacterium bovis BCG. *Vaccine* **32**, 1362-1367
130. Romanowski R., Culbert R., Alekmade S., Medellin-Pena M.J., Bugarski D., Milanovic A., and A., M. (2017) MYCOBACTERIUM CELL WALL FRACTION IMMUNOSTIMULANT (AMPLIMUNE™) EFFICACY IN THE REDUCTION OF THE SEVERITY OF ETEC INDUCED DIARRHEA IN NEONATAL CALVES. *Acta Veterinaria-Beograd* **67**, 222-237
131. Yang, J., Mao, M., Zhang, S., Li, H., Jiang, Z., Cao, G., Cao, D., Wang, X., and Zhang, L. (2012) Innate defense regulator peptide synergizes with CpG ODN for enhanced innate intestinal immune responses in neonate piglets. *Int Immunopharmacol* **12**, 415-424
132. Turanek, J., Kasna, A., Koudela, B., Ledvina, M., and Miller, A. D. (2005) Stimulation of innate immunity in newborn kids against Cryptosporidium parvum infection-challenge by intranasal/per-oral administration of liposomal formulation of N-L18-norAbu-GMDP adjuvant. *Parasitology* **131**, 601-608
133. Liu, Y., Zhang, J., Wang, S., Guo, Y., He, T., and Zhou, R. (2019) A Novel Adjuvant "Sublancin" Enhances Immune Response in Specific Pathogen-Free Broiler Chickens Inoculated with Newcastle Disease Vaccine. *Journal of immunology research* **2019**, 1016567
134. Mattioli, G. A., Rosa, D. E., Turic, E., Testa, J. A., Lizarraga, R. M., and Fazzio, L. E. (2019) Effect of Injectable Copper and Zinc Supplementation on Weight, Hematological Parameters, and Immune Response in Pre-weaning Beef Calves. *Biological trace element research* **189**, 456-462

Chapter 7

RNA Vaccines for Infectious Diseases of Animals: A Future Trend

Laurel J. Gershwin*

Dept. of Pathology, Microbiology and Immunology, School of Veterinary Medicine, University of California Davis, Davis, CA 95616, USA

*Corresponding author: ljgershwin@ucdavis.edu

DOI: https://doi.org/10.21775/9781913652593.07

Abstract

Future trends for veterinary vaccines must include mRNA vaccines. Vaccines consisting of mRNA that encodes a protective antigen from the pathogen of interest offer a new and exciting approach to development of veterinary immunoprophylaxis. The technique has several advantages over attenuated, inactivated, subunit, and viral vectored vaccines. Once the sequence of a candidate antigen is known a mRNA can be made. Cell culture and viral propagation is not a requirement for vaccine production. Thus these vaccines can be manufactured rapidly and once developed are not costly. These characteristics make mRNA vaccines a strong choice for development of newly emerging viruses for species of veterinary importance. Indeed at least one candidate vaccine for the COVID-19 currently in clinical trials is a mRNA vaccine. Herein the published work on candidate mRNA vaccines for human and zoonotic disease is described, including mechanisms of action and choices for delivery of the mRNA are discussed.

Introduction

The science of vaccinology has had a huge impact on control and eradication of disease in both human and veterinary species. The earliest successful vaccines were developed centuries ago; for example, in 1796 by Jenner (cowpox as a vaccine for smallpox) and in the 1800's by Pasteur (fowl cholera and rabies). These early vaccines were based on either attenuation or inactivation of a pathogen so that the modified form or inactivated pathogen would elicit immunization but not disease. Despite many success stories in veterinary medicine, these classical methods of vaccine production have not always been adequate. Inactivated pathogen vaccines generally induce only humoral immunity and they require the use of adjuvants and usually booster immunizations. As the science of vaccinology has grown over the decades additional types of vaccines have been developed: subunit vaccines and viral vectored recombinant vaccines been developed using molecular biology techniques that make it possible to introduce to the patient only the important antigens required to stimulate protective immunity, without unneeded pieces of the pathogen. Taken one step further, DNA vaccines were developed which contained only the genetic material coding for an antigen administered as a component of a plasmid. For DNA vaccines to work uptake into host cells and transfer through the nuclear membrane must precede production of the protein antigens coded for by the DNA incorporated into the plasmid and once expressed these antigens are then presented to the immune system to stimulate immunity. Several DNA vaccines have been developed for use in veterinary species, but overall the efficacy has been somewhat disappointing, particularly in application to infectious diseases of livestock.

The newest innovation in vaccine development is the mRNA vaccine (Pardi et al., 2018). mRNA vaccines have the advantage of stimulating innate immunity by engaging pattern recognition receptors, such as TLR3. TLR7, and TLR8, which are present on endosomes in the cytoplasm. RNA vaccines do not need to enter the cell nucleus and are thus able to initiate immune responses in the cytoplasm. This is an advantage of mRNA vaccines over DNA based vaccines which must enter the nucleus to instigate production of the antigenic proteins for which they code. Entry of DNA into the host nucleus has been shown to be inefficient, perhaps explaining why the immune response is less robust than anticipated. RNA vaccines have been shown to initiate antigen expression resulting in robust humoral and cellular immune responses without the need for an adjuvant. Because the development of RNA vaccines does not require cell culture or propagation of viral vectors, these vaccines can be rapidly developed and manufactured in the event of a pandemic (as we are currently seeing in the human population due to a unique coronavirus, COVID19).

Types of RNA Vaccines

The initial steps required to develop an RNA vaccine involve first sequencing the genome of the pathogen then using the sequence of the pathogen to design vaccine candidates. There are two types of RNA vaccines described (Maruggi et al., 2019). These are: conventional mRNA vaccines and self-amplifying mRNA vaccines (SAM). The process of making a mRNA vaccine requires knowing which viral proteins/ antigens are candidates for inducing a protective immune response. Once the antigen is selected its sequence is synthesized it is cloned into a DNA template plasmid for production of the mRNA in vitro. The mRNA is then used to immunize the host where in vivo translation of the mRNA produces the antigen. This basic process can be further modified by altering codon usage and/or by sequence modification as shown for development of mRNA as therapeutics (Thess et al., 2015). As seen in figure 1 the conventional mRNA vaccine consists of the coding sequence for the

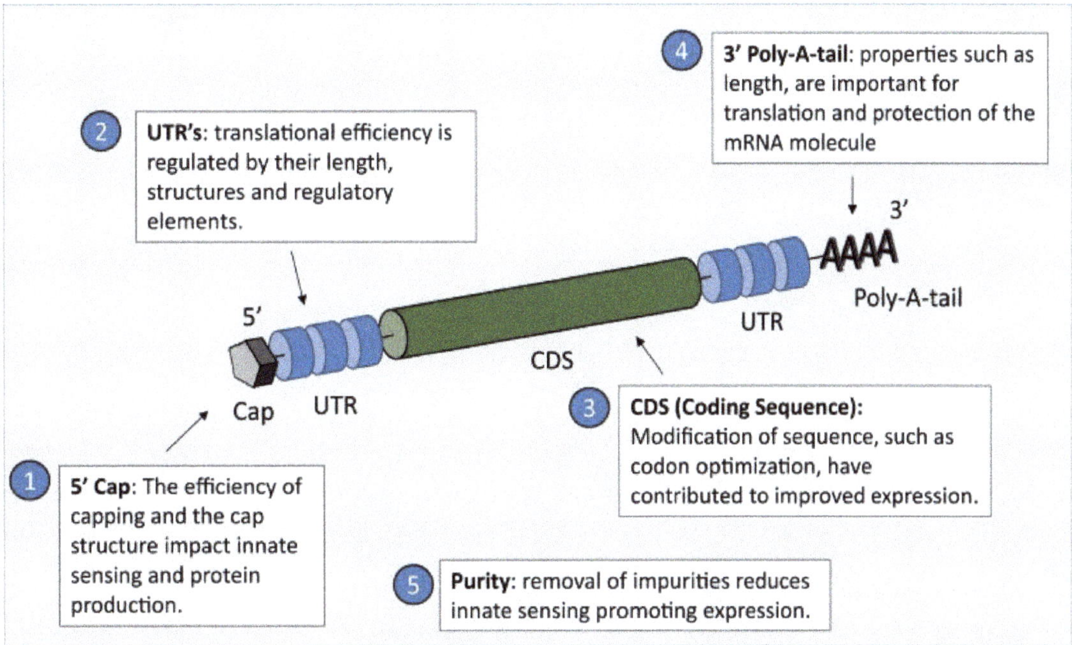

Figure 1. The structure of the mRNA construct includes the 5' cap (1), the UTR (2), the coding sequence which contains the genetic information for the desired protein to be expressed (3), and the poly-A-tail (4). This figure is courtesy of npj Vaccines (2020) 5:11; https://doi.org/10.1038/s41541-020-0159-8.

antigen flanked by a 5'UTR on one side and a 3' UTR on the other side, a terminal 5' cap and a terminal 3'poly A tail (Maruggi et al., 2019; Jackson et al., 2020).

The non-replicating mRNA vaccine is taken into the cell by endocytosis and ultimately escapes the endosome to the cytosol where the constructs are translated into protein by the ribosomes. The protein product can then undergo post-translational modification and peptides are then presented with MHC molecules on the cell surface (figure 2 top). Protein products can also become transmembrane

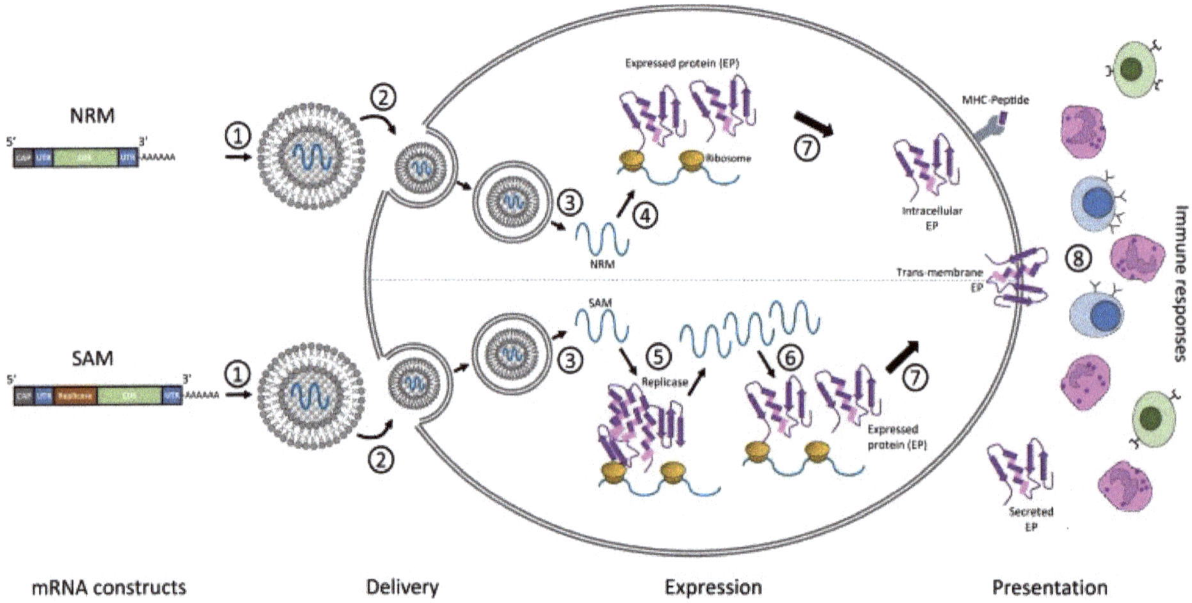

Figure 2. Top: Non-replicating mRNA is encapsulated in LNP (1). It is taken up into the cell by endocytosis (2). mRNA is released from the endosome into the cytosol (3). Ribosomes translate the mRNA into the protein encoded by the mRNA (4). Post translational modification occurs and the expressed proteins are either processed for membrane expression on MHC molecules, secreted or affixed in the cell membrane as transmembrane proteins (7).
Bottom: The self-amplifying (SAM) mRNAs in LNP enter in a similar manner as the non-replicating mRNA (1-3). The SAM are immediately translated by ribosomes to create the replicase required for self-amplification of the mRNA (5). The self-amplified mRNA is translated by ribosomes to produce the protein and the proteins receive post-translational modification (6). Expressed proteins are then generated as secreted, trans-membrane or intracellular proteins (7). The proteins that were coded for by the mRNA are then available for interaction with cells of the innate and acquired immune systems. This figure is courtesy of npj Vaccines (2020) 5:11; https://doi.org/10.1038/s41541-020-0159-8.

proteins or secreted proteins. These proteins are detected by innate and adaptive immune cells.

The second type of mRNA vaccine is the self-amplifying mRNA (SAM) vaccine. This type is usually made from the genome of a positive strand RNA virus, such as alpha viruses. Besides the antigen coding sequence it contains the coding sequences for nonstructural proteins from the virus; these are important for amplification of the RNA in the intracellular compartment and enhances antigen expression. Positive strand RNA viruses are able to act as mRNA directly when they enter the cell cytoplasm. The non-structural proteins become the RNA dependent polymerase and facilitate transcription of the mRNA into protein. (figure 2 bottom). Studies have shown that the self-amplifying mRNA vaccine can induce higher levels of antigen for a longer period of time than the conventional mRNA vaccine. This resulted in a more potent immune response from the self-amplifying mRNA vaccine (Vogel et al., 2018).

There are advantages and disadvantages of each of these mRNA vaccine types. As described by Maruggi et al (Maruggi et al., 2019) both types of mRNA vaccines are non-infectious and do not integrate into host DNA. They are able to be scaled up for production rapidly due to the lack of need to propagate virus in tissue culture. The in vivo expression stimulates the innate immune system. When comparing the two types, it appears that lower mRNA doses can be used in the self-replicating mRNA vaccine, while the conventional type requires more RNA. The conventional type mRNA vaccine often utilizes nucleoside base modification to improve potency of the vaccine. Sequence optimization is also used to enhance protein expression and immunogenicity, often by adding more G-C 's (Maruggi et al.,2019). The increased size of the self- amplifying mRNA vaccine means that it is possible to insert more than one coding gene for antigen and still have a relatively low dose vaccine.

Delivery of mRNA vaccines

The ability to deliver mRNA into cells is critical to the use of mRNA as a vaccine. The degradability of unprotected RNA is well known as RNA degrading enzymes are plentiful. In addition mRNA is unlikely to pass through the cell membrane in a naked form, due to charge and size. For this reason there are several potential methods being evaluated for delivery of mRNA vaccines into cells. Methods of delivery include: lipid nanoparticles, electroporation, cationic miscelles, silica based nanoparticles, as well as viral vectors. Delivery of mRNA into the cytoplasm of cells by gene gun has been used as an alternative to encapsulated mRNA for delivery of mRNA vaccines (Qiu et al., 1996). In early studies with DNA vaccines gene gun administration was a primary means of delivery. However, the more recent studies that have been done with mRNA vaccines are focused on parenteral administration with standard injection syringes and needles.

Lipid nanoparticles

Lipid nanoparticles have been shown to have low toxicity and good immune stimulation (Zhang et al., 2019). Lipid nanoparticles (LNP) have been used in studies focused on delivery of siRNA into cells for therapeutic purposes (Kanasty et al., 2013). Early work with lipid nanoparticles focused on drug delivery to several body compartments by topical, oral, and parenteral administration (Ghasemiyeh and Mohammadi-Samani, 2018). Lipid nanoparticles are made from phospholipids and are solid at both body and room temperature. They range in size from 50 to 1000 nm. When enclosed in the LNP mRNA is protected from enzymatic degradation and entry into the cell is facilitated. This process is accomplished by the process of endocytosis with ultimate release of the mRNA from endosomes into the cytosol, where translation occurs. LNPs are often made with cationic outer membranes to facilitate entry into the cell. The major delivery methods of mRNA have been described (Pardi et al., 2018). These include: naked mRNA, electroporation, complexed with protamine, cationic nano-emulsion, modified dendrimer nanoparticle, protamine liposome, cationic polymer, cationic polymer liposome, polysaccharide particle, cationic lipid nanoparticle, cationic lipid, cholesterol nanoparticle, cationic lipid cholesterol, PEG nanoparticle.

The route by which these mRNA carrying LNPs are administered is also important. While the focus of this chapter is on the use of mRNA vaccines to prevent infectious diseases, there has been considerable work done with mRNA vaccines for cancer prophylaxis and therapy, providing some excellent protocols (Sayour et al., 2018). The delivery routes that have been tested include intravenous, intraperitoneal, intramuscular, intradermal and subcutaneous. Expression of protein and duration of effect were superior for intramuscular and intradermal routes in mice (Pardi et al., 2017).

Silica nanoparticles

Silica nanoparticles are another method of delivery for mRNA vaccines. These silica nanoparticles are designed to have a spiky surface that traps and protects the mRNA from nuclease enzymes. The particles enter the cell by endocytosis. Once released into the cytoplasm the mRNA induces both humoral and cellular responses (Theobald, 2020). It is reported that silica nanoparticles produce less inflammatory reaction when entering the cells than the LNPs do (Theobald, 2020).

Induction of immune Responses by RNA Vaccines

The success of a mRNA vaccine requires engagement of both the innate and acquired immune system. When mRNA from a viral pathogen enters the host the exogenous RNA is sensed by members of the toll like receptor (TLR) family that are germline encoded pattern recognition receptors: endosomal TLR3, 7, 8 and RIG1-like receptors in the cytosol. These receptors recognize ds RNA (TLR3) and ss RNA (TLR7 and 8). The RIG1 and the MDA-5 are both pattern recognition receptors present in the differentiated tissues; so it is likely that mRNA injected into muscle would initially encounter these receptors (Deering et al., 2014).The result of these interactions is activation of type 1 interferon gene expression with resultant transcription of interferon stimulated genes. These innate immune interactions initiate inflammatory cytokine production, which is important in getting recruitment of antigen presenting cells for initiation of the acquired immune response. One downside of this exuberant response is enhanced inflammation; however, nucleoside modifications during the in vitro preparation of the vaccine mRNA have rendered these molecules less likely to cause severe inflammatory reactions. This process involves incorporation of chemically modified nucleosides, for example, pseudouridine or 5-methyl-cytidine into the mRNA transcript (Pardi et al., 2018).

The migration of dendritic cells to local lymph nodes initiates the acquired arm of the immune system. In these secondary lymphoid structures T follicular helper cells interact with dendritic cells presenting antigen; and germinal center B cells are stimulated to differentiate, ultimately producing plasma cells making antibody against the expressed antigen. The follicular helper cells are important for further development of the immune response through affinity maturation and somatic hypermutation (Pardi et al., 2018). Thus mRNA vaccines must induce this population of cells for a solid and lasting acquired immune response. A strong follicular helper T cell response was shown for both HIV and Influenza viruses in separate studies that used a single immunization with mRNA in LNP coding for viral antigens in a mouse model (Pardi et al., 2018). CD8 T cell responses have also been described in a variety of mouse studies (Deering et al., 2014).

Examples of mRNA vaccines in Experimental Studies and Clinical Trials

The earliest studies on mRNA vaccines were for expression of tumor antigens for stimulation of an immune response against tumors. Intradermal injections of mRNA coding for tumor associated antigens showed promise for induction of an immune response in human patients with malignant melanoma (Rittig et al., 2011).

More recently the use of LNP as a delivery vehicle for mRNA vaccines has been applied to a variety of viruses including: Influenza, Zika, Ebola, Rabies, and HIV (Maruggi et al., 2019). For all of these mRNA vaccines formulated in LNPs an immune response was shown in mice or other laboratory animals.

A mRNA vaccine trial for Zika virus was successful in both mouse and primate models. A single dose intradermal vaccine of a LNP vaccine was administered; mice received 30 ug of a nucleoside modified mRNA with the gene for Zika virus prM-E and this produced neutralizing antibody titers. E protein specific IgG persisted through 20 weeks post immunization. Immunized mice and placebo injected mice were challenged IV at 2 and 20 weeks with virulent Zika virus and all vaccinated mice were protected from viremia. Rhesus macaque monkeys were vaccinated with 600ug, 200ug or 50ug of the same vaccine construct ID, resulting in both E protein specific IgG and neutralizing antibody titers from all three doses. The monkeys were challenged subcutaneously with virulent Zika virus and monitored for viremia. The lowest dose (50 ug) was protective at the 5 weeks challenge timepoint. This study demonstrates in two animal species effectiveness of the LNP mRNA vaccine for immune stimulation and protection from challenge infection (Pardi et al., 2017).

A mRNA vaccine study performed for immunization against Ebola virus showed promise. For rapidly emerging diseases like Ebola the ability to make the mRNA vaccine rapidly and scale it up for mass distribution makes it an attractive format. In this study guinea pigs were immunized with the Ebola glycoprotein (GP) mRNA vaccine by the intramuscular route twice 3 weeks apart. They were then challenged

21 days later by the intraperitoneal route with guinea pig adapted Ebola virus. They developed GP specific IgG by 2 weeks after the initial vaccine, which was boosted by the live virus exposure. Neutralizing antibody titers were similarly induced and boosted by challenge exposure. Placebo immunized guinea pigs developed severe disease after challenge, while the immunized animals failed to show clinical signs of illness (Meyer et al., 2017).

Currently (2020) there are a variety of phase 1 trials underway for mRNA vaccines for viral diseases of humans. These include: Respiratory syncytial virus, Rabies, Influenza, Cytomegalovirus, and Zika virus (Jackson et al., 2020). Development of a pandemic from emergence of a new corona virus has sparked rapid development of mRNA vaccines targeting the spike protein of the virus; multiple human vaccine trials are currently underway.

mRNA vaccine trials for diseases of importance in veterinary medicine

Rabies virus

Rabies is one of the diseases most commonly vaccinated for in veterinary medicine. Rabies virus is highly neurotropic and causes a fatal disease in humans, generally after exposure of a human from a bite by an infected animal. In the United States rabies vaccination is required for all dogs and strongly advised for pet cats and horses. In some states where Rabies is endemic in wildlife the threat of rabies virus infection from animals such as skunks and bats is very high. In developing countries where rabies vaccination is not commonly done, the threat of a fatal human or animal infection from a bite wound is great.

Although the original vaccine for rabies was developed by Louis Pasteur in 1885 by attenuation of the virus in rabbit spinal cord, the current vaccine for rabies in veterinary species is killed rabies virus in adjuvant. In 2000 a canary pox vectored rabies vaccine containing the immunogenic glycoprotein became available for cats (Jas et al., 2012). A vaccinia vectored rabies glycoprotein vaccine and was licensed for racoons and coyotes, administered in bait, for these wild populations of potential rabies carriers (Maki, 2017). There is no live attenuated rabies virus vaccine currently in use.

Rabies is a highly fatal zoonotic infectious disease and vaccination against this disease is required for individuals with a high likelihood of exposure (veterinarians, animal care takers, etc.). There is a need for an economic non-replicating vaccine that can induce both neutralizing antibodies and a T cell response. Schnee et al developed a mRNA vaccine encoding the rabies virus glycoprotein G and tested it in mice and in domestic pigs (Schnee et al 2016). This mRNA vaccine was based on the Pasteur strain and did not include chemically modified nucleotides; it was complexed with protamine and was administered by the intradermal route in both mice and pigs. Immune parameters were evaluated. T cell analysis using intracellular cytokine staining by flow cytometry and Elispot showed induction of antigen specific responses in both CD4+ and CD8+ T cell populations in mice; Interferonγ and TNFα were shown by intracellular staining. Studies in domestic adult and newborn piglets showed seroconversion with protective levels of virus neutralizing antibody titers. In both mouse and pig studies the animals were protected from virulent challenge (Schnee et al., 2016).

Influenza virus

Another zoonotic disease that has been addressed with mRNA vaccine technology is Influenza. Influenza virus is an important pathogen for humans, horses, dogs and other species, which can harbor and transmit the virus. The influenza virus is a negative strand RNA virus that has a segmented genome. The major surface glycoproteins, hemagglutinin (HA) and neuraminidase (N), are important immune targets. However, due to the lack of a proof- reading function of the polymerase proteins it is common to have mutations in these genes, leading to altered antigenic proteins. This is called antigenic drift and is responsible for the yearly differences in strains that circulate in the human population. When two or more strains of Influenza undergo exchange/reassortment of RNA segments antigenic shift occurs. This phenomenon can occur between avian and mammalian influenza viruses or between strains that primarily infect different species of mammals.

The ability of influenza virus to alter its H and N antigens and thus avert immune responses previously stimulated makes it necessary for new vaccines to be made on a yearly basis for the human population. Traditional vaccines require growth of virus in tissue culture or eggs prior to inactivation for production of inactivated vaccines. The mRNA vaccine has the promise of allowing more rapid response to changing antigenic epitopes, with the ability to mass produce the vaccine without requirements for growth of cells in culture and infection with virus for vaccine production.

It is well recognized that the human-animal interface is an important mechanism for transmission of Influenza virus. Influenza readily crosses the species barrier to infect a new host; examples include avian influenza and swine influenza. The emergence of influenza outbreaks in dogs in 2015 was traced to influenza in horses. Close contact between poultry and humans has caused epidemic waves in China. Greater detail on past emergence of epidemic influenza outbreaks due to cross-species transmission is reviewed (Rajao and Perez., 2018).

All human influenza pandemics in the last 100 years have come from viruses with gene segments from animal viruses (Taubenberger and Morens, 2009). This is due to the close contact with the human population and animal species, such as poultry and swine. Hence the ability to rapidly identify new emergent strains and to develop protective vaccines is critical for prevention of wide- spread disease.

The mRNA vaccine technology is well suited to the problems inherent in providing protection from Influenza to populations of humans and animals. A number of studies have been done to evaluate the potential for usage of mRNA Influenza vaccines in humans and other species. Both non-replicating mRNA vaccines and SAM vaccines have been developed and tested for Influenza. Target antigens include the hemagglutinin (HA), nucleoprotein (NP), and the matrix protein 1 (MP1). In one study protective neutralizing antibody responses to HA in mice were achieved with intramuscular doses of 0.1 or 1ug in a SAM LNP vaccine format. An NP and M1 SAM vaccine format was shown to induce potent antigen specific T cell responses in mice (Scorza et al., 2018).

Demonstration of an effective immune response in mice, ferrets and nonhuman primates against Influenza viruses (H10N8 and H7N9) was achieved using mRNA vaccine encoding the appropriate H and N proteins. The vaccines were delivered using lipid nanoparticles. Neutralizing antibody titers as well as protection from virulent challenge were seen in mice and ferrets. Phase 1 Immunogenicity studies of the modified mRNA LNP vaccine showed excellent antibody titers with little or no adverse responses (Bahl et al., 2017).

Conclusions

There have been a variety of mRNA vaccine studies performed using laboratory species as well as non-human primates. Early studies focused on cancer vaccines and subsequently the focus has included infectious diseases of humans and those pathogens of zoonotic importance. Encouraging results from such studies demonstrated induction of strong protective immunity including both humoral and cell mediated responses with little or no adverse consequences. Methods for delivery of the mRNA to the host cells include many choices, including the lipid nanoparticle, which appears to be an excellent candidate for immunization of animals. When a rapid response is needed to protect a population of animals from a newly emerging pathogen the mRNA vaccine will be a robust candidate for initiation of a balanced and effective immune response.

Future Trends

mRNA vaccines are a strong contender for the most important veterinary vaccine platform of the future. They can be made rapidly, effectively, with little or no side effects, and are far more economic for widespread use than traditional vaccines. As newly emerging viruses confront livestock and companion animals by entering our country or already prevalent pathogens require alternative vaccines that can induce strong humoral and cellular responses, the RNA vaccine will likely become a staple for vaccine production enterprises. Future applications will also include oncology and possibly transplantation medicine.

References

Bahl K, Senn JJ, Yuzhakov O, et al. (2017). Preclinical and Clinical Demonstration of Immunogenicity by mRNA Vaccines against H10N8 and H7N9 Influenza Viruses. *Mol Ther.* 25(6):1316-1327. doi:10.1016/j.ymthe.2017.03.035

Deering RP., Kommareddky S., Ulmer JB., Brito LA., & Geall AJ. (2014). Nucleic acid vaccines: prospects for non-viral delivery of mRNA vaccines. Expert Opin. Drug Deliv. 11(6)885-898.

Ghasemiyeh P, Mohammadi-Samani S. (2018). Solid lipid nanoparticles and nanostructured lipid carriers as novel drug delivery systems: applications, advantages and disadvantages. *Res Pharm Sci.* 2018;13(4):288-303. doi: 10.4103/1735-5362.235156

Guevara ML, Persano S, Persano F. Lipid-Based Vectors for Therapeutic mRNA-Based Anti-Cancer Vaccines. (2019). *Curr Pharm Des.* 25(13):1443-1454. doi: 10.2174/1381612825666190619150221

Jackson NAC, Kester KE, Casimiro D, Gurunathan S, DeRosa F. (2020). The promise of mRNA vaccines: a biotech and industrial perspective. *NPJ Vaccines.* 5:11. doi:10.1038/s41541-020-0159-8

Jas D, Coupier C, Toulemonde CE, Guigal PM, Poulet H. (2012) Three-year duration of immunity in cats vaccinated with a canarypox-vectored recombinant rabies virus vaccine. *Vaccine.* 30(49):6991-6996. doi:10.1016/j.vaccine.2012.09.068

Kanasty R, Dorkin JR, Vegas A, Anderson D. (2013). Delivery materials for siRNA therapeutics. *Nat Mater.* 12(11):967-977. doi:10.1038/nmat3765

Kulkarni JA, Witzigmann D, Chen S, Cullis PR, van der Meel R. (2019). Lipid Nanoparticle Technology for Clinical Translation of siRNA Therapeutics. *Acc Chem Res.* 52(9):2435-2444. doi:10.1021/acs.accounts.9b00368

Maki J, Guiot AL, Aubert M, et al. (2017). Oral vaccination of wildlife using a vaccinia-rabies-glycoprotein recombinant virus vaccine (RABORAL V-RG®): a global review. *Vet Res.* 48(1):57. doi:10.1186/s13567-017-0459-9

Maruggi G, Zhang C, Li J, Ulmer JB, Yu D. (2019). mRNA as a Transformative Technology for Vaccine Development to Control Infectious Diseases. *Mol Ther.* 27(4):757-772. doi:10.1016/j.ymthe.2019.01.020

Meyer M, Huang E, Yuzhakov O, Ramanathan P, Ciaramella G, Bukreyev A. (2018). Modified mRNA-Based Vaccines Elicit Robust Immune Responses and Protect Guinea Pigs From Ebola Virus Disease. *J Infect Dis.* 217(3):451-455. doi:10.1093/infdis/jix592.

Oberli MA, Reichmuth AM, Dorkin JR, et al. (2017). Lipid Nanoparticle Assisted mRNA Delivery for Potent Cancer Immunotherapy. *Nano Lett.* 17(3):1326-1335. doi:10.1021/acs.nanolett.6b03329

Pardi, N., Hogan, M. J., Pelc, R. S., Muramatsu, H., Andersen, H., DeMaso, C. R., Dowd, K. A., Sutherland, L. L., Scearce, R. M., Parks, R., Wagner, W., Granados, A., Greenhouse, J., Walker, M., Willis, E., Yu, J. S., McGee, C. E.,

Sempowski, G. D., Mui, B. L., Tam, Y. K., ... Weissman, D. (2017). Zika virus protection by a single low-dose nucleoside-modified mRNA vaccination. *Nature*, *543*(7644), 248–251. https://doi.org/10.1038/nature21428

Pardi, N., Hogan, M. J., Naradikian, M. S., Parkhouse, K., Cain, D. W., Jones, L., Moody, M. A., Verkerke, H. P., Myles, A., Willis, E., LaBranche, C. C., Montefiori, D. C., Lobby, J. L., Saunders, K. O., Liao, H. X., Korber, B. T., Sutherland, L. L.,

Scearce, R. M., Hraber, P. T., Tombácz, I., ... Weissman, D. (2018). Nucleoside-modified mRNA vaccines induce potent T follicular helper and germinal center B cell responses. *The Journal of experimental medicine, 215*(6), 1571–1588. https://doi.org/10.1084/jem.20171450

Pardi, N., Hogan, M. J., Porter, F. W., & Weissman, D. (2018). mRNA vaccines - a new era in vaccinology. *Nature reviews. Drug discovery, 17*(4), 261–279. https://doi.org/10.1038/nrd.2017.243.

Qiu P, Ziegelhoffer P, Sun J, Yang NS. (1996). Gene gun delivery of mRNA in situ results in efficient transgene expression and genetic immunization. *Gene Ther.* 3(3):262-268.

Rajão DS, Pérez DR. (2018). Universal Vaccines and Vaccine Platforms to Protect against Influenza Viruses in Humans and Agriculture. *Front Microbiol.* 9:123. doi: 10.3389/fmicb.2018.00123

Sayour EJ, Mendez-Gomez HR, Mitchell DA. (2018). Cancer Vaccine Immunotherapy with RNA-Loaded Liposomes. *Int J Mol Sci.* 19(10):2890. doi: 10.3390/ijms19102890

Schnee M, Vogel AB, Voss D, et al. (2016). An mRNA Vaccine Encoding Rabies Virus Glycoprotein Induces Protection against Lethal Infection in Mice and Correlates of Protection in Adult and Newborn Pigs. *PLoS Negl Trop Dis.* 10(6):e0004746. doi:10.1371/journal.pntd.0004746

Scorza FB, Pardi N. (2018). New Kids on the Block: RNA-Based Influenza Virus Vaccines. *Vaccines (Basel).*;6(2):20. doi:10.3390/vaccines6020020

Taubenberger JK, Morens DM. (2019). The 1918 Influenza Pandemic and Its Legacy. *Cold Spring Harb Perspect Med.* ;a038695. doi:10.1101/cshperspect.a038695

Theobald N. (2020). Emerging vaccine delivery systems for COVID-19: Functionalised silica nanoparticles offer a potentially safe and effective alternative delivery system for DNA/RNA vaccines and may be useful in the hunt for a COVID-19 vaccine. *Drug discovery today,* S1359-6446(20)30248-8. Advance online publication. https://doi.org/10.1016/j.drudis.2020.06.020

Ulmer JB, Mason PW, Geall A, Mandl CW. (2012). RNA-based vaccines. *Vaccine.* 30(30):4414-4418. doi:10.1016/j.vaccine.2012.04.060.

Vogel AB, Lambert L, Kinnear E, et al. (2018). Self-Amplifying RNA Vaccines Give Equivalent Protection against Influenza to mRNA Vaccines but at Much Lower Doses. *Mol Ther*. 26(2):446-455. doi:10.1016/j.ymthe.2017.11.017

Zhang, C., Maruggi, G., Shan, H., & Li, J. (2019). Advances in mRNA Vaccines for Infectious Diseases. *Frontiers in immunology*, *10*, 594. https://doi.org/10.3389/fimmu.2019.00594.

www.ingramcontent.com/pod-product-compliance
Lightning Source LLC
LaVergne TN
LVHW070852210926
848482LV00016B/920